MW01537812

Solutions Manual

for

BIOCHEMISTRY
A FOUNDATION

Peck Ritter
Eastern Washington University

Brooks/Cole Publishing Company

I(T)P® An International Thomson Publishing Company

Pacific Grove • Albany • Bonn • Boston • Cincinnati • Detroit
London • Madrid • Melbourne • Mexico City • New York • Paris
San Francisco • Singapore • Tokyo • Toronto • Washington

Art for 15.5: Based on *Principles of Biochemistry,* Second Edition, by A. L. Lehninger, D. L. Nelson, and M. M. Cox, Worth Publishers, 1993.
Art for 16.2: Copyright Irving Geis.

Assistant Editor: *Elizabeth Barelli Rammel*
Cover Design: *Lisa Berman*
Cover Illustration: *Ken Eward/Biografx*
Editorial Associate: *Beth Wilbur*
Marketing Team: *Connie Jirovsky & Romy Fineroff*
Production Editor: *Mary Vezilich*
Printing and Binding: *Malloy Lithographing*

I(T)P The ITP logo is a registered trademark under license.

For more information, contact:

BROOKS/COLE PUBLISHING COMPANY
511 Forest Lodge Road
Pacific Grove, CA 93950
USA

International Thomson Editores
Campos Eliseos 385, Piso 7
Col. Polanco
11560 México D. F. México

International Thomson Publishing Europe
Berkshire House 168-173
High Holborn
London WC1V 7AA
England

International Thomson Publishing GmbH
Königswinterer Strasse 418
53227 Bonn
Germany

Thomas Nelson Australia
102 Dodds Street
South Melbourne, 3205
Victoria, Australia

International Thomson Publishing Asia
221 Henderson Road
#05-10 Henderson Building
Singapore 0315

Nelson Canada
1120 Birchmount Road
Scarborough, Ontario
Canada M1K 5G4

International Thomson Publishing Japan
Hirakawacho Kyowa Building, 3F
2-2-1 Hirakawacho
Chiyoda-ku, Tokyo 102
Japan

Printed in the United States of America

5   4   3   2   1

ISBN 0-534-33867-4

Contents

1

Introduction

1.1   See text. Each term in bolded type is defined, described, etc. at that point in the text where it is first encountered. Most bolded terms are worth remembering. Try placing those terms emphasized by your instructor on flash cards for study purposes.

1.2   The natural sciences appear to be governed by fixed laws that cannot be modified by any living organisms, including man. Alas, no such laws govern the social sciences. The problems of social scientists are compounded by the tendency of their human subjects to be less than completely open and honest. In the natural sciences, we always obtain the same results when we test hypotheses with carefully designed and controlled experiments. This is not always the case in the social sciences.

1.3   c, f and g. Scientific laws are simply statements about how we believe nature operates. Although it may seem unlikely that existing laws will change, future observations may lead to changes in our present beliefs. This does not mean that the way nature operates has changed or will change.

1.4   The similarities are summarized in Exhibit 1.2. The major differences are listed in Exhibit 1.3.

1.5   a, d, e and f.

Water, Acids, Bases and Buffers

2.1

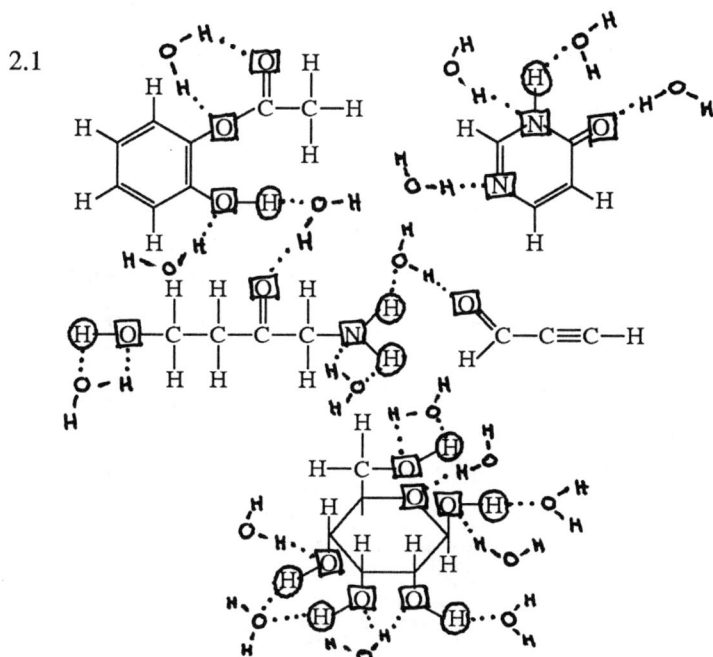

2.2 Van der Waals, hydrophobic forces, dipole-dipole interactions (other than H bonds), H bonds, ion-dipole interactions, ion pairs (Table 2.5). Ion pairs, ion-dipole interactions, dipole-dipole interactions, and H bonds are electrostatic interactions (Table 2.4).

2.3 Any class of organic compound with a charged functional group will tend to form micelles when its charged functional group is attached to a long hydrocarbon chain or some other

3

long nonpolar chain. These classes of organic compounds include salts of thiols, salts of amines, salts of phosphoric acid esters and quaternary ammonium salts (R_4N^+).

2.4   Ionic bonds. Positively charged protonated amino groups will be attracted to negatively charged dissociated carboxyl groups. In the absence of a polar solvent, the bond energy will be substantial (Table 2.5).

2.5   Esters, alcohols, amides, anhydrides, amines, ethers, acetals, hemiacetals, aldehydes, ketones and others. Some of these classes contain oxygens and/or nitrogens that can H bond to the hydrogens in water but lack hydrogens capable of H bonding.

2.6

	Sample	$[H^+]*$	$[OH^-]*$	$pH^\dagger$	$POH^\dagger$	Acidic, Basic or Neutral?
	Example. Pure water	1×10^{-7}	1×10^{-7}	7	7	neutral
1.	0.020 M HNO_3	0.020 M	5.0×10^{-13} M	1.7	12.3	acidic
2.	0.18 M KOH	5.5×10^{-14} M	0.18 M	13.3	0.7	basic
3.	0.18 M H_2CO_3 plus 0.11 M HCO_3^-	7.3×10^{-7} M	1.4×10^{-8} M	6.1	7.9	acidic
4.	0.37 M $H_2PO_4^-$ plus 0.02 M H_3PO_4	3.8×10^{-4} M	2.6×10^{-11} M	3.4	10.6	acidic
5.	0.90 M NH_4^+ plus 0.08 M NH_3	6.3×10^{-9} M	1.6×10^{-6} M	8.2	5.8	basic
6.	0.25 M NaCl	1.0×10^{-7}	1.0×10^{-7}	7.0	7.0	neutral
7.	0.25 M NaCl plus 0.25 M HCl	0.25 M	4.0×10^{-14} M	0.6	13.4	acidic
8.	0.15 M lactate plus 0.75 M lactic acid	6.9×10^{-4} M	1.4×10^{-11} M	3.2	10.8	acidic
9.	1.0 M ethanol	1.0×10^{-7} M	1.0×10^{-7} M	7.0	7.0	neutral
10.	1.2 M NaOH	8.3×10^{-15}	1.2 M	14.08	−0.08	basic

*All answers rounded off to two significant figures.
$\dagger$With the exception of the last row, all answers expressed to one tenth of a unit.

Applicable equations and definitions:
$K_w = [H^+][OH^-] = 10^{-14}$
$pH = -\log[H^+]$
$pOH = -\log[OH^-]$
Acid Solution – $[H^+] > [OH^-]$
Basic Solution – $[OH^-] > [H^+]$
Neutral Solution – $[H^+] = [OH^-]$
$pH = pK_a + \log[(CB)/(WA)]$

Sample #1: HNO_3 is a strong acid (dissociates completely). In a 0.020 M solution, $[H^+]$ = 0.020 M; $[OH^-] = 10^{-14}/0.020$ M.

Samples #2 and #10: Both KOH and NaOH are strong bases (dissociate completely). The molarity of each solution equals the $[OH^-]$; $[H^+] = 10^{-14}/[OH^-]$.

Samples #3, #4, #5 and #8: Each is a buffer solution in which you are given the concentration of the weak acid and its conjugate base. pH = pK_a + log ([CB]/[WA]). In sample #3, for example:

$$pH = 6.35 + \log (0.11/0.18) = 6.35 + (-0.214) = 6.136 \cong 6.1$$

$$pH = -\log [H^+]; \quad [H^+] = \text{antilog} (-pH) = 7.3 \times 10^{-7} \text{ M}$$

Samples #6 and #7: Since NaCl is a salt whose ionic components are neither acids nor bases, its presence will have no impact on the [H$^+$] in a solution. A pure NaCl solution should have the same pH as the water in which it was dissolved. Pure water has a pH = 7. The HCl (a strong acid) in solution #7 will lead to [H$^+$] = 0.25 M; [OH$^-$] = 10^{-14}/0.25 = 4.0 × 10^{-14} M.

Solution #9: Since ethanol has no significant acidic or basic properties (it has very little tendency to either accept or donate H$^+$), its presence has no significant impact on [H$^+$]. A pure 1.0 M ethanol solution has a pH = 7.

2.7   1. $H_2SO_4 + 2KOH \rightleftarrows 2H_2O + K_2SO_4$

2. $HNO_3 + NaHCO_3 \rightleftarrows H_2CO_3 + NaNO_3$

3. $CH_3NH_2 + HCl \rightleftarrows CH_3NH_3^+ + Cl^-$

4. $K_2HPO_4 + NaOH \rightleftarrows H_2O + NaK_2PO_4$

5. $K_2HPO_4 + HCl \rightleftarrows KCl + KH_2PO_4$

6. $CH_3CH_2SH + NaOH \rightleftarrows H_2O + CH_3CH_2SNa$

7. $CH_3OPO_3H^- + NaOH \rightleftarrows H_2O + CH_3OPO_3Na^-$

8. $CH_3OPO_3H^- + HCl \rightleftarrows CH_3OPO_3H_2 + Cl^-$

9. $CH_3CH_2CH_2COOH + KOH \rightleftarrows H_2O + CH_3CH_2CH_2COOK$

10. $CH_3CH_2COOK + HNO_3 \rightleftarrows CH_3CH_2COOH + KNO_3$

The first reaction involves a strong acid and a strong base. In reactions 2, 3, 5, 8 and 10, a strong acid reacts with the conjugate base of a weak acid and converts that conjugate base to its weak acid. In reactions 4, 6, 7 and 9, a strong base reacts with a weak acid to generate water plus the conjugate base of the weak acid. To solve problems of this type, one must recognize the acids and bases involved.

Each of the equations includes one or more spectator ions, usually not shown as ions since they are associated with a counterion in an ionic compound.

Notice that K_2HPO_4 and $CH_3OPO_3H^-$ can function as either acids or bases. Such compounds are called amphoteric compounds.

2.8   a) $HCOO^-$

        b) $(CH_3)_2AsO_2^-$

        c) $C_6H_5COO^-$

        d) $(CH_3)_2NH$

        e) F^-

        f) HS^-

        g) CN^-

In each case, the conjugate base contains one less H^+ than its weak acid. As a consequence, the net charge on the weak acid is one unit more positive than the net charge associated with its conjugate base.

2.9   a) $K_{eq} = \dfrac{[H^+][HCOO^-]}{[HCOOH]}$

        b) $K_{eq} = \dfrac{[H^+][(CH_3)_2AsO_2^-]}{[(CH_3)_2AsO_2H]}$

        c) $K_{eq} = \dfrac{[H^+][C_6H_5COO^-]}{[C_6H_5COOH]}$

        d) $K_{eq} = \dfrac{[H^+][(CH_3)_2NH]}{[(CH_3)_2NH_2^+]}$

        e) $K_{eq} = \dfrac{[H^+][F^-]}{[HF]}$

        f) $K_{eq} = \dfrac{[H^+][HS^-]}{[H_2S]}$

        g) $K_{eq} = \dfrac{[H^+][CN^-]}{[HCN]}$

2.10   d (weakest), g, f, b, c, a, e (strongest),

Since an equilibrium constant is a measure of the extent to which a reaction occurs, the larger the K_a, the stronger the acid. Since $pK_a = -\log K_a$, the smaller the pK_a, the stronger the acid.

2.11   a) 2.75–4.75

   b) 5.27–7.27

   c) 3.20–5.20

   d) 9.80–11.8

   e) 2.18–4.18

   f) 5.89–7.89

   g) 8.31–10.31

Buffer range = $pK_a \pm 1$. A solution must have a significant amount of both a weak acid and its conjugate base before it is able to consume a significant amount of both added acid and added base. This requirement is only met when the pH is relatively close to the pK_a (Table 2.10).

2.12   a) 1.78×10^{-4}

   b) 5.37×10^{-7}

   c) 6.31×10^{-5}

   d) 1.58×10^{-11}

   e) 6.61×10^{-4}

   f) 1.29×10^{-7}

   g) 4.90×10^{-10}

$pK_a = -\log K_a$; by the definition of antilog, $K_a = $ antilog $(-pK_a)$

2.13   There are many possible answers. One can, for example, write the formula for any three carboxylic acids, phenols, thiols or protonated amines whose formulas have not been given in this text. All members of these classes of compounds are weak acids. Examples: $CH_3(CH_2)_{10}COOH$, BrC_6H_4OH, $HSCH_2CH_2SH$, $CH_3(CH_2)_5NH_3^+$.

2.14   i) $H^+ + F^- \rightleftarrows HF$; $OH^- + HF \rightleftarrows H_2O + F^-$

   ii) $H^+ + HSO_3^- \rightleftarrows H_2SO_3$; $OH^- + H_2SO_3 \rightleftarrows H_2O + HSO_3^-$

   iii) $H^+ + SH^- \rightleftarrows H_2S$; $OH^- + H_2S \rightleftarrows H_2O + HS^-$

   iv) $H^+ + CHCl_2COO^- \rightleftarrows CHCl_2COOH$; $OH^- + CHCl_2COOH \rightleftarrows H_2O + CHCl_2COO^-$

   v) $H^+ + C_6H_5COO^- \rightleftarrows C_6H_5COOH$; $OH^- + C_6H_5COOH \rightleftarrows H_2O + C_6H_5COO^-$

   vi) $H^+ + CH_3CHOHCH_2S^- \rightleftarrows CH_3CHOHCH_2SH$;
       $OH^- + CH_3CHOHCH_2SH \rightleftarrows H_2O + CH_3CHOHCH_2S^-$

vii) $H^+ + CH_3CH = CHCH_2NH_2 \rightleftarrows CH_3CH = CHCH_2NH_3^+$;
    $OH^- + CH_3CH = CHCH_2NH_3^+ \rightleftarrows H_2O + CH_3CH = CHCH_2NH_2$

viii) $H^+ + RO^- \rightleftarrows ROH$; $OH^- + ROH \rightleftarrows H_2O + RO^-$ where R = $— C_6H_4OOCCH_3$

The HNO_3 converts the conjugate base in each buffer pair to its weak acid. NO_3^- is a spectator ion. The KOH converts the weak acid in each buffer pair to its conjugate base by removing H^+ to form water. K^+ is a spectator ion.

2.15

	pH 1	pH 6	pH 10	pH 14
Succinic acid	a) 0	b) –1.5	c) –2	d) –2
Fructose 6-phosphate	a) –0.5	b) –1.5	c) –2	d) –2
5-Hydroxytryptamine	a) +1	b) +1	c) +0.5	d) –1
Onithine	a) +2	b) +1	c) 0	d) –1

A functional group that can exist in both a weak acid and conjugate base form is assigned the charge of its weak acid form if the pH is more than 0.5 below the pK_a of this weak acid. It is assigned the charge of its conjugate base form when the pH is 0.5 or more above the pK_a of its weak acid form. Such a functional group is assigned a charge that is the average of the charges on its two forms whenever the pH is within 0.5 of the pK_a of its weak acid form. The net charge on a molecule (to the nearest half of a charge unit) is obtained by adding all of the charges assigned to its individual functional groups. The rules for assigning charges were derived, as described in this chapter, from the Henderson-Hasselbalch equation ($pH = pK_a + \log ([CB]/[WA])$).

2.16   pH10   i)

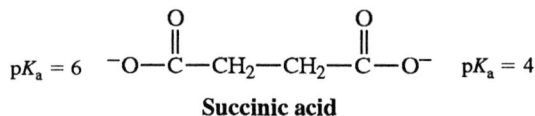

Succinic acid

(ii)

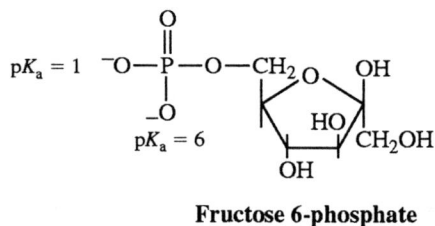

Fructose 6-phosphate

(iii)

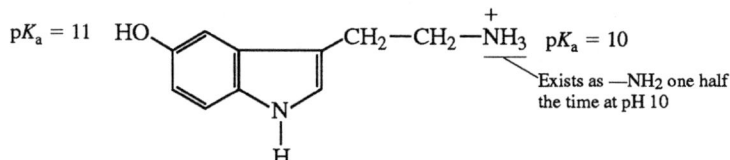

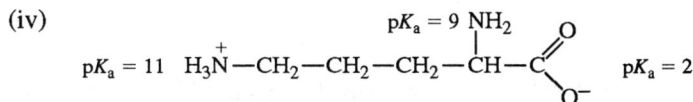

(iv)

$pK_a = 9$ NH$_2$

$pK_a = 11$ H$_3\overset{+}{N}$—CH$_2$—CH$_2$—CH$_2$—CH—C $\quad pK_a = 2$

O / O$^-$

Ornithine

pH 1   i)

O

$pK_a = 6$ HO—C—CH$_2$—CH$_2$—C—OH $\quad pK_a = 4$

O

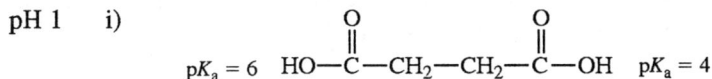

Succinic acid

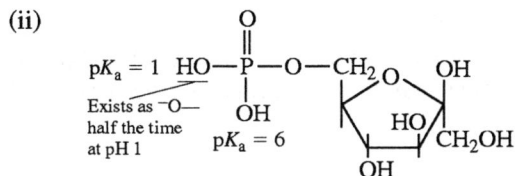

(ii)

O

$pK_a = 1$ HO—P—O—CH$_2$ O   OH

Exists as $^-$O—
half the time   OH
at pH 1 $pK_a = 6$ HO   CH$_2$OH

OH

Fructose 6-phosphate

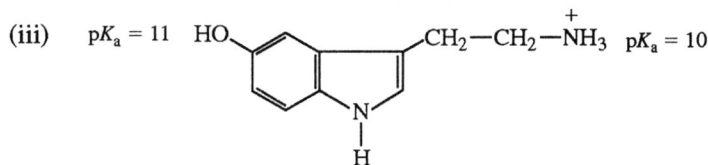

(iii) $pK_a = 11$ HO

CH$_2$—CH$_2$—$\overset{+}{N}$H$_3$ $pK_a = 10$

N
|
H

5-Hydroxytryptamine

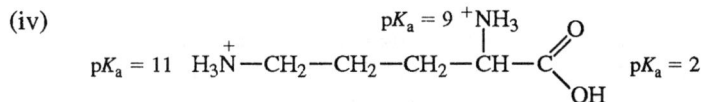

(iv)

$pK_a = 9$ $^+$NH$_3$

$pK_a = 11$ H$_3\overset{+}{N}$—CH$_2$—CH$_2$—CH$_2$—CH—C $\quad pK_a = 2$

O / OH

Ornithine

A weak acid functional group will primarily exist in its protonated form at pH's below its pK_a and primarily exist in its dissociated form at pH's above its pK_a. When the pH = pK_a, a functional group will exist in its protonated form half the time and in its dissociated form half the time. These generalizations are based on the mathematical interpretation of an algebraically-revised form of the Henderson-Hasselbalch equation: [CB]/[WA] = antilog (pH $-pK_a$).

2.17   4/10

Rearrangement of the Henderson Hasselbalch equation yields: [CB]/[WA] = antilog (pH – pK_a). The pK_a for $H_2PO_4^-$ is 7.20 (Table 2.8). At pH 6.8, [CB]/[WA] = antilog (6.8 – 7.2) = 0.40/1. To arrive at a whole number ratio, we multiply both the numerator and the denominator by 10: (0.40 × 10)/(1 × 10) = 4/10. For every 4 molecules that exist as HPO_4^{2-}, approximately 10 will exist as $H_2PO_4^-$.

2.18   pH = 4.76

The HCl reacts with 1 mol of the CH_3COONa (sodium acetate, a source of the conjugate base of acetic acid) converting it to CH_3COOH (acetic acid). Since 1 mol of CH_3COONa remains, both acetic acid and its conjugate base are present at a concentration of 1 mol/L. pH = pK_a + log ([CB]/[WA]) = 4.76 + log 1/1 = 4.76.

2.19   Acidosis

As CO_2 accumulates in blood, H_2CO_3 accumulates as well; $CO_2 + H_2O \rightleftarrows H_2CO_3$. The partial dissociation of this weak acid ($H_2CO_3 \rightleftarrows H^+ + HCO_3^-$) increases the [$H^+$] and decreases the pH.

2.20   7.83

$$pH = pK_a + \log ([CB]/[WA]) = 6.35 + \log ([HCO_3^-]/[H_2CO_3])$$
$$= 6.35 + \log 30 = 6.35 + 1.48 = 7.83.$$

2.21   Venous blood has a higher CO_2 concentration; CO_2 from the oxidation of organic fuel molecules continuously enters the blood as it flows from the lungs back to the lungs. $CO_2 + H_2O \rightleftarrows H_2CO_3 \rightleftarrows H^+ + HCO_3^-$; also see Exercise 2.19

2.22   4.37

Since the sodium acetate dissociates completely to yield Na^+ and CH_3COO^- (the conjugate base of acetic acid), pH = pK_a + log ([CB]/[WA]) = 4.76 + log (0.15/0.37) = 4.76 + (–0.39) = 4.37.

2.23   pH = 13.3

The HCl will neutralize 1 mol of the NaOH ($H^+ + OH^- \rightleftarrows H_2O$) leaving 0.2 mol of unreacted NaOH in 1 L. Since NaOH is a strong base that dissociated completely (NaOH $\rightleftarrows Na^+ + OH^-$), [$OH^-$] = 0.2 M. Since [$H^+$][$OH^-$] = 10^{-14}, [H^+] = $10^{-14}/0.2 = 5 \times 10^{-14}$. pH = –log [$H^+$] = 13.3.

2.24   See text. Each of these terms is defined or described at the point where it is introduced.

2.25   a) pH = 4.7;   b) pH = 5.4;   c) pH = 3.7;   d) pH = 2.0;   e) pH = 3.4

In each case, one uses the following approach:

$$K_a = \frac{[H^+][CB]}{[WA]}$$

Assume [CB] = [H$^+$]; one H$^+$ and one conjugate base (CB) are formed for each molecule of weak acid (WA) that dissociates. This assumption neglects the small amount of H$^+$ that exists due to the dissociation of water.

Assume that the final [WA] equals the initial concentration of weak acid. This assumption neglects the drop in [WA] that results from the dissociation of a small fraction of the weak acid.

Plugging these assumed values into the equilibrium constant expression one obtains:

$$K_a = \frac{[H^+]^2}{[WA]}$$

$$[H^+]^2 = K_a[WA]; \quad [H^+] = \text{square root of } K_a[WA]$$

$$pH = -\log[H^+]$$

a) $[H^+] = \sqrt{K_a[WA]}$ = square root [$(4.90 \times 10^{-10})(0.72)$] = 1.9×10^{-5} M; pH = 4.7

b) $[H^+] = \sqrt{(1.58 \times 10^{-11})(1.2)}$ = 4.4×10^{-6} M; pH = 5.4

c) $[H^+] = \sqrt{(1.29 \times 10^{-7})(0.36)}$ = 2.2×10^{-4} M; pH = 3.7

d) $[H^+] = \sqrt{(1.78 \times 10^{-4})(0.48)}$ = 9.2×10^{-3} M; pH = 2.0

e) $[H^+] = \sqrt{(5.37 \times 10^{-7})(0.25)}$ = 3.7×10^{-4} M; pH = 3.4

2.26   a) HCOOH;   b) $(CH_3)_2NH_2^+$;   c) HF;   d) H_2S;   e) HCN

The conjugate acid of a weak base is that acid which is formed when the base accepts an H$^+$.

2.27   a) $K_b = 5.62 \times 10^{-11}$; $pK_b = 10.3$

b) $K_b = 6.33 \times 10^{-4}$; $pK_b = 3.20$

c) $K_b = 1.51 \times 10^{-11}$; $pK_b = 10.8$

d) $K_b = 7.75 \times 10^{-8}$; $pK_b = 7.11$

e) $K_b = 2.04 \times 10^{-5}$; $pK_b = 4.69$

K_b's and pK_b's can be calculated from the pK_a's given in Exercise 2.8 by using the following equations and any one of multiple calculation paths:

$$pK_a = -\log K_a;\ pK_b = -\log K_b$$
$$K_a K_b = K_w = 10^{-14};\ pK_a + pK_b = 14$$

a) $K_b = K_w/K_a = 10^{-14}/1.78 \times 10^{-4} = 5.62 \times 10^{-11}$

Note: K_a = antilog $(-pK_a)$.

$pK_b = -\log K_b = -\log (5.62 \times 10^{-11}) = 10.3$

b) $K_b = 10^{-14}/(1.58 \times 10^{-11}) = 6.33 \times 10^{-4};\ pK_b = -\log (6.33 \times 10^{-4}) = 3.20$

c) $K_b = 10^{-14}/(6.61 \times 10^{-4}) = 1.51 \times 10^{-11};\ pK_b = -\log (1.51 \times 10^{-11}) = 10.8$

d) $K_b = 10^{-14}/(1.29 \times 10^{-7}) = 7.75 \times 10^{-8};\ pK_b = -\log (7.75 \times 10^{-8}) = 7.11$

e) $K_b = 10^{-14}/(4.90 \times 10^{-10}) = 2.04 \times 10^{-5};\ pK_b = -\log (2.04 \times 10^{-5}) = 4.69$

2.28   a) $K_b = \dfrac{[HCOOH][OH^-]}{[HCOO^-]}$

b) $K_b = \dfrac{[(CH_3)_2NH_2^+][OH^-]}{[(CH_3)_2NH]}$

c) $K_b = \dfrac{[HF][OH^-]}{[F^-]}$

d) $K_b = \dfrac{[H_2S][OH^-]}{[HS^-]}$

e) $K_b = \dfrac{[HCN][OH^-]}{[CN^-]}$

2.29   c (weakest), a, d, e, b (strongest)

The larger the K_b, the stronger the base. K_b's were calculated in Exercise 2.27.

3

Amino Acids, Peptides and
Three Laboratory Techniques

3.1   The structures of all the common protein amino acids are given in Figure 3.1.

3.2   a) pH 1; none   pH 7; none   pH 14; 1 and 5:   b) 4 and 5   c) 2, 3 and 6   d) 1 and 5   e) 4, alcohol; all, carboxylic acids; 1, thiol; all, amines; 5, phenol   f) 1, 2, 3 and 6 (both 1 and 2 are borderline polar/nonpolar—see Table 3.5 and associated text discussion)

   a) Only the side chains of 1 (a thiol) and 6 (a phenol) are capable of existing in a negatively charged form (R—S$^-$ and Ar—O$^-$, respectively). These side chains will have a charge of –1 (to the nearest half charge unit) whenever the pH is 0.5 or more above their pK$_a$'s (Section 2.7).

   b) Most side chains containing a N, O or F are capable of H bonding. When one of these highly electronegative atoms is present, it usually creates a charge separation (within a bond) great enough to allow H bonding (Section 2.1). According to some authors, sulfhydryl groups (—SH) also participate, albeit poorly, in H bonding.

   c) Saturated hydrocarbons are relatively nonreactive and possess no functional groups. Functional groups, in general, are reactive centers (groupings of atoms) that contain carbon to carbon multiple bonds or elements in addition to H and C.

   d) There are three weak acid functional groups in the depicted side chains: a thiol functional group (1; called a sulfhydryl group), a phenol functional group (5; called a hydroxyl group) and an alcohol functional group (4; also called a hydroxyl group). Phenols and thiols are weak acids. Technically, alcohols are as well, but they are so weak (pK$_a$'s usually 14 or above) that they are often considered to be nonacids from a practical standpoint.

13

e) Phenols have an —OH attached to an aromatic carbon. Alcohols have an —OH attached to an isolated aliphatic carbon. Thiols, considered sulfur analog of alcohols, have an —SH attached to an isolated carbon. Carboxylic acids are characterized by the presence of a carboxyl group (—COOH). Amines contain a nitrogen bonded to one or more carbons (H_2NR, HNR_1R_2, or $NR_1R_2R_3$).

f) A side chain must have one or more significantly polar bonds to be polar. This requirement immediately excludes compounds 2, 3 and 6. Compound 2 is, however, classified as borderline polar/nonpolar, because it lacks significant nonpolar character as well as significant polar character. Compound 1 is also ranked borderline polar/nonpolar for reasons described in Section 3.2. The other compounds contain one or more polar bonds that do not cancel. See the answer to Problem 3.9 for a review of polarity.

3.3   a.

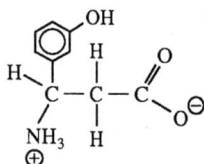

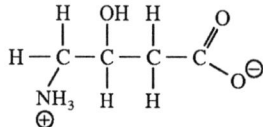

There are many possible answers. In each case, an amino group must be attached to a carbon that is β to a carboxyl group and an —OH must be attached to an aromatic ring.

b.

There are many possible answers. In each case, an amino group must be attached to a carbon that is γ to a carboxyl group and an —OH must be attached to an aliphatic carbon that is not part of a larger functional group.

c.

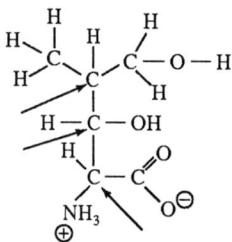

There are many possible answers. The chiral centers in this example are identified with arrows.

d.

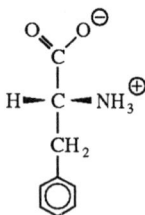

To illustrate the configuration about the α-carbon of an amino acid, one must follow specific conventions. The carboxyl group should be written above and the side chain written below the α-carbon. Both are implied to be behind the α-carbon. An amino group written to the right of the α-carbon identifies a D-configuration while an amino group to the left signifies an L-configuration.

e.

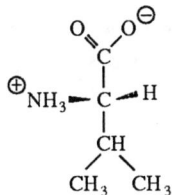

3.4

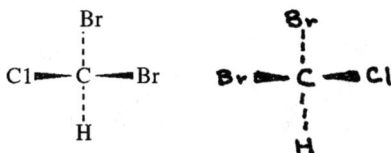

Not enantiomers since mirror image can be superimposed on original compound . Original compound is not chiral (it is achiral).

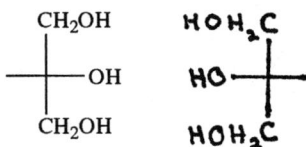

Not enantiomers. Mirror images are superimposable. Original compound is achrial.

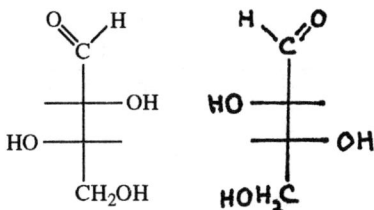

Enantiomers. Mirror images are **not** superimposable.

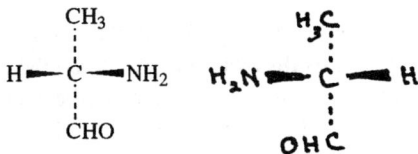

Enantiomers. Mirror images are **not** superimposable.

3.5   No. Amino acid analysis is designed to determine the relative ratio of amino acids in a peptide. It provides no information about the absolute number of copies of an amino acid in the peptide being analyzed. Ala-Gly and Ala-Gly-Ala-Gly would both yield the same results.

3.6   a)–f)

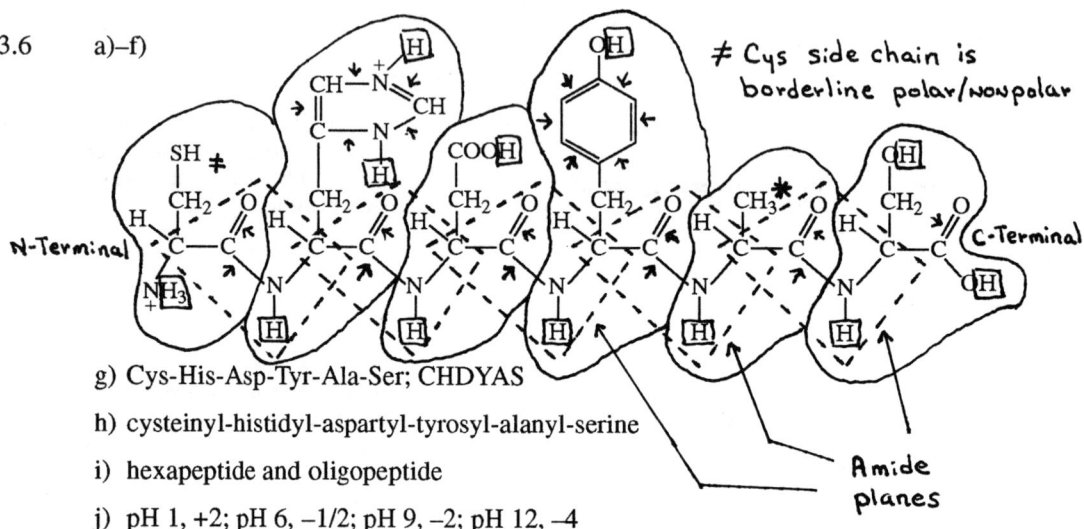

≠ Cys side chain is borderline polar/nonpolar

Amide planes

g) Cys-His-Asp-Tyr-Ala-Ser; CHDYAS

h) cysteinyl-histidyl-aspartyl-tyrosyl-alanyl-serine

i) hexapeptide and oligopeptide

j) pH 1, +2; pH 6, –1/2; pH 9, –2; pH 12, –4

	N-terminal –NH$_2$	Cys Side Chain	His Side Chain	Asp Side Chain	Tyr Side Chain	C-Terminal —COOH	Net Charge
	Charge (to nearest 1/2 unit)						
pH 1	+1	0	+1	0	0	0	+2
pH 6	+1	0	+1/2	–1	0	–1	–1/2
pH 9	+1	–1	0	–1	0	–1	–2
pH 12	0	–1	0	–1	–1	–1	–4

A weak acid functional group is assigned the charge of its protonated form if the pH is 0.5 or more below its pK$_a$ and assigned the charge of its dissociated form when the pH is 0.5 or more above its pK$_a$. It is assigned a charge equal to the average of the charges on its protonated and dissociated forms whenever the pH is within 0.5 of its pK$_a$. The basis for these rules are examined in Section 2.7.

k) pH 1, cathode;   pH 6, 9 and 12, anode

Opposite charges attract. If its net charge is negative, a molecule will move towards the anode (+). Net positively-charged molecules move towards the cathode (–).

l) pH 1, cation exchange;   pH 6, 9 and 12, anion exchange

Opposite charges attract. Since cation exchange columns have negatively-charged stationary phases, they bind positively charged ions (cations). Negative anions do not bind to negative cation exchangers. Similarly, cations do not bind to positively-charged anion exchangers that bind and exchange anions.

3.7   a) Oxytocin—a pituitary hormone that stimulates lactation and the contraction of uterine muscles. It is commonly used to induce labor during childbirth. It also plays a role in the functioning of the central nervous system.

b) Endorphins—naturally occurring opiate peptides that function as neurotransmitters.

c) Vasopressin—a pituitary hormone that helps regulate water balance and blood pressure. It also functions as a neurotransmitter, and it may help regulate memory consolidation and sexual behavior.

d) Glutamate—a building block for proteins, a neurotransmitter and a common food additive (flavor enhancer). Leads to the Chinese-restaurant syndrome in sensitive individuals.

3.8   Pentapeptides: Ala-Ala-Thr-Leu-Leu; Ala-Ala-Leu-Thr-Leu; Ala-Thr-Ala-Leu-Leu; Leu-Ala-Leu-Thr-Ala; plus many other possible answers.

Decapeptide: Ala-Ala-Ala-Ala-Thr-Thr-Leu-Leu-Leu-Leu and Ala-Thr-Ala-Leu-Leu-Thr-Ala-Leu-Ala-Leu are among the large number of possible answers.
The four pentapeptides would be very difficult to separate with electrophoresis since they all have virtually the same size, charge and shape. With classical electrophoresis, separation is primarily determined by differences in charge to size ratio.

3.9   pH 1

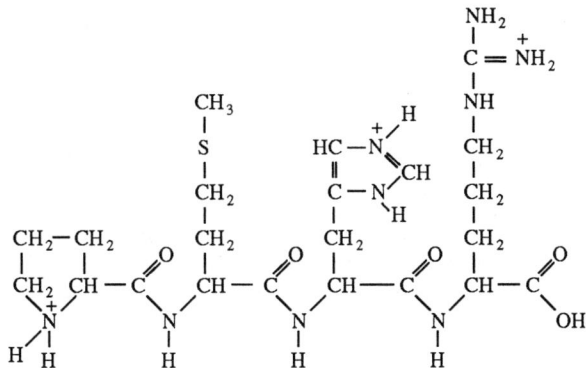

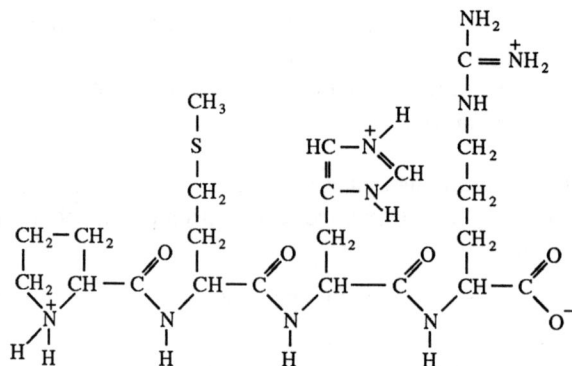

pH 5

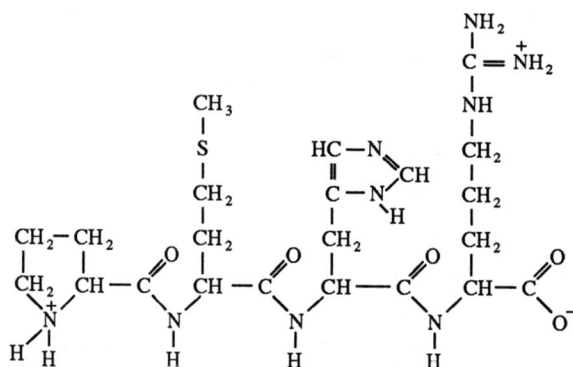

pH 8

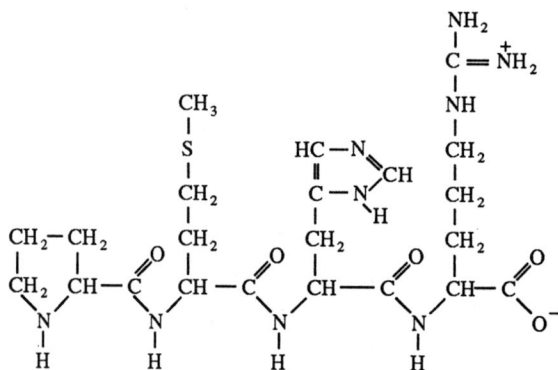

pH 12

pH 1, binds cation exchanger;   pH 5, binds cation exchanger

pH 8, binds cation exchanger;   pH 12, binds anion exchanger

Cation exchangers (–) bind and exchange cations (+) and compounds with a net positive charge while anion exchangers (+) bind and exchange anions (–) and compounds with a net negative charge. The net charge on the given tetrapeptide is clearly positive at pH 1, 5 and 8. At pH 12, the guanidinium group of arginine is partially dissociated while all of the other functional groups are almost totally dissociated. Since the partial dissociation of the arginine side chain reduces its charge contribution to less than +1, it does not totally cancel the –1 charge associated with the C-terminal carboxyl: a net negative charge results. Precise charge calculations are not needed; one must simply determine whether the net charge is positive or negative.

A weak acid functional group exists predominantly in its protonated form if the pH is below its pK_a and mainly in its dissociated form when the pH is above its pK_a. The two forms exist in equal concentrations whenever $pH = pK_a$ (Section 2.7).

3.10   True. Since dipeptides and polypeptides differ greatly in size, gel filtration beads with appropriate-sized pores will accommodate the smaller dipeptides but exclude polypeptides. When the dipeptides enter the pores, their migration through the column will be retarded relative to that of the polypeptides.

3.11 $\varepsilon_{220} = 1.08 \times 10^5$ L mol^{-1} cm^{-1}; One would need an absorbance measurement at 345 nm to calculate ε_{345}.

$$A = \varepsilon bc \qquad \varepsilon = \frac{A}{bc} = \frac{1.08}{(1 \text{ cm})(1 \times 10^{-5} \text{ M})} = 1.08 \times 10^5 \text{ L mol}^{-1} \text{ cm}^{-1}$$

Pathlength is assumed to be 1 cm (considered standard) unless specified otherwise.

3.12 1.22×10^{-4} M

$$A = \varepsilon bc \qquad c = \frac{A}{\varepsilon b} = \frac{0.56}{(2.3 \times 10^3 \text{ L mol}^{-1} \text{ cm}^{-1})(2 \text{ cm})} = 1.22 \times 10^{-4} \text{ M}$$

3.13 $A_{340} = 0.8 \qquad A_{560} = 1.2$

Each drug absorbs the same amount of radiation that it would if it were the only absorbing compound in the solution. Total absorbance is, therefore, equal to the sum of the absorbance of a 2×10^{-5} M Drug T solution and a 4×10^{-5} M Drug P solution:
$A = \varepsilon bc$

$$A_{340} = (2 \times 10^4)(1)(2 \times 10^{-5}) + (1 \times 10^4)(1)(4 \times 10^{-5}) = 0.4 + 0.4 = 0.8$$

$$A_{560} = (0)(1)(2 \times 10^{-5}) + (3 \times 10^4)(1)(4 \times 10^{-5}) = 0 + 1.2 = 1.2$$

Pathlength is once again assumed to be 1 cm.

3.14 0.25 cm

$$A = \varepsilon bc \qquad b = \frac{A}{\varepsilon c} = \frac{1}{(4 \times 10^2 \text{ L mol}^{-1} \text{ cm}^{-1})(0.01 \text{ M})} = 0.25 \text{ cm}$$

The pathlength is in centimeters because ε is in L mol^{-1} cm^{-1}.

3.15

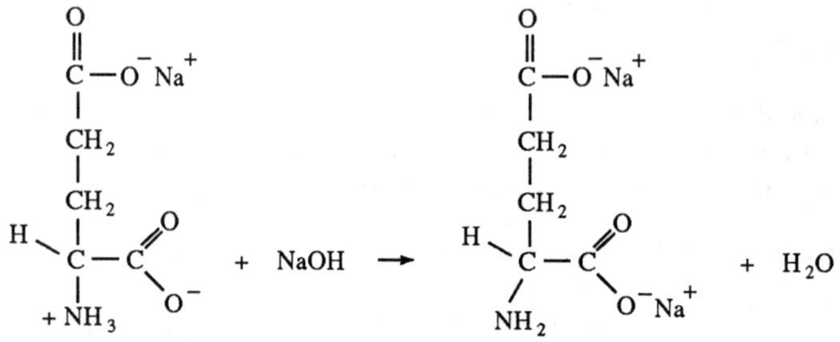

In monosodium glutamate, the protonated α-amino group is the counter ion to the α-carboxylate ion. When NaOH removes the H$^+$ from the —NH$_3$$^+$, Na$^+$ becomes the new counterion.

3.16 Formaldehyde (official name is methanal) and formic acid (official name is methanoic acid).

Primary alcohols are oxidized to aldehydes which can, in turn, be oxidized to carboxylic acids.

3.17 $^{30}_{15}$P

3.18 False. By definition, two atoms must have the same atomic number and be atoms of the same element in order to be isotopes.

3.19 half-life = 31.8 hr

$$\text{half-life} = \frac{0.301 \times 6 \text{ hr}}{\log(6200/5440)} = \frac{1.81}{0.0568}$$

3.20 If one assumes that 6.02 = 6.00, 24 hr = 4 half-lives and 1 µCi will decay to 1/16 µCi = **0.0625 µCi**: $1 \rightarrow 1/2 \rightarrow 1/4 \rightarrow 1/8 \rightarrow 1/16$

For a more precise answer, you can use the formula provided in Exercise 3.19:

$$\text{half-life} = \frac{0.301 t}{\log(N_o/N)}$$

An algebraic rearrangement leads to:

$$\log(N_o/N) = \frac{0.301t}{\text{half-life}} = \frac{0.301 \times 24 \text{ hr}}{6.02} = 1.20$$

$(N_o/N) = \text{antilog } 1.20 = 15.8$

When $N_o = 1 \ \mu\text{Ci}, \frac{1}{N} = 15.8$ and $N = 0.0633 \ \mu\text{Ci}$

3.21   Tuberculosis (TB) is now rare, although it is making a comeback. When TB is rare, the benefits of routine chest X-rays no longer justify the costs and the radiation risks associated with the X-rays. In diagnosis and treatment decisions, one is commonly faced with a risk-benefit analysis.

3.22   Geiger counters directly detect and count electric pulses (avalanches of electrons) created by radiation-induced ionizations.

Scintillation counters directly detect and count radiation-induced flashes of light.

3.23 5.54×10^{-8} Ci; 2.05×10^3 Bq

Since counting efficiency $= \dfrac{\text{cpm}}{\text{dpm}} \times 100,$

$\text{dpm} = \dfrac{\text{cpm} \times 100}{\text{counting efficiency}} = 37,000 \times 100/30 = 123,000$

$123,000 \text{ dpm} \times \dfrac{1 \text{ dps}}{60 \text{ dpm}} \times \dfrac{1 \text{ Ci}}{3.7 \times 10^{10} \text{ dps}} = 5.54 \times 10^{-8} \text{ Ci}$

$5.54 \times 10^{-8} \text{ Ci} \times \dfrac{3.7 \times 10^{10} \text{ Bq}}{1 \text{ Ci}} = 2.05 \times 10^3 \text{ Bq}$

3.24   Administer a precise amount of drug X, and then remove and count samples of blood at several times after the administration. The rate at which ^{14}C accumulates in blood is a good measure of the rate at which the drug enters the blood if: a) the drug remains intact within the digestive system so that all of the ^{14}C entering the blood resides in Drug X; and b) no significant amount of Drug X leaves the blood during the time period where measurements are made. It is unlikely that such an experiment would be approved for humans because there is a radiation risk with no associated benefits.

3.25   No. Compound F may be rapidly converted to another compound that is accumulated by cancer cells. To determine if the ^{14}C resides in intact compound F, one must isolate and identify the radioactive compounds in cancer cells.

3.26   See text. **Bolded** terms are described, defined and/or discussed at those sites where they appear.

4

Proteins

4.1   The protein is soluble in pure water (is an albumin) and has a nucleic acid prosthetic group (is a nucleoprotein). It has multiple ligand binding sites (required for cooperative binding) and it is probably an oligomer (cooperative binding usually entails communication between multiple subunits within an oligomer).

4.2   See Table 4.1.

4.3   See Table 4.1. In addition to the specific proteins listed in Table 4.1, the text discusses keratin (the structural protein in hair) and myoglobin (stores O_2 in tissues). Brief reference is also made to several specific enzymes (protein catalysts) and some additional proteins.

4.4   True, by definition. For a compound to be classified as a protein it must have at least one polypeptide chain. However, not all polypeptide chains constitute proteins. Normally, a polypeptide must be naturally-occurring, must possess 50 or more amino acid residues and must exist in a unique conformation under physiological conditions before it is considered a protein. Polypeptides within oligomeric proteins are usually considered protein subunits rather than intact proteins.

4.5   Asp-Phe-Val, Asp-Val-Phe, Phe-Asp-Val, Phe-Val-Asp, Val-Phe-Asp, and Val-Asp-Phe.

There are 24 possible tetrapeptides containing one copy each of Ser, Met, Arg, and Tyr. The answer is the same for any set of four different amino acids. Although this answer can fairly rapidly be arrived at with a trial and error approach, a statistician knows that the answer is 4 factorial = $4 \times 3 \times 2 \times 1 = 24$. Similarly, the number of different peptides that contain one copy of each of the 20 protein amino acids is 20 factorial = $20 \times 19 \times 18 \times 17 \times 16 \times \cdots 2 \times 1 = 2.43 \times 10^{18}$, an extremely large number. If one allows the same amino acid to occur in multiple copies, it can be shown that there are 20^n different peptides that each contain n amino acid residues. Thus, there are 20^{100} possible polypeptides that each contain 100 amino acid residues. The potential polypeptide diversity is truly astronomical.

4.6   Asp-Gly-Ser-Lys-Met-Phe-Glu-Met-His-Ile-Lys-Ala-Thr-Met-Pro-Tyr-Arg-Gln-Val

For discussion purposes, number the peptides in Group 1 from 1 to 4, and then assign the letters A through D to the peptides in Group 2. The Lys-Met sequence in Peptide C, indicates that Peptide 2 must have been joined (within Peptide P) to the N-terminal end of Peptide 1. The Ile-Lys-Ala-Thr sequence in Peptide D leads to the conclusion that Peptide 3 was originally joined to the C-terminal side of Peptide 1. Similarly, the sequence of Peptide B places Peptide 4 to the C-terminal side of Peptide 3.

4.7   No. Amino acid analysis provides absolutely no information about the ordering of the amino acids along the polypeptide chain. All proteins containing the same ratio of amino acids yield exactly the same results.

4.8   The F helix contains 9 residues, the H helix 24. A small region of the H helix appears to contact the F helix. Since helix-helix contacts occur predominantly within the core of the protein, rather than on its surface, one would expect the contacts to be primarily nonpolar in nature—nonpolar in, polar out (the general rule for the folding of globular proteins in an aqueous environment).

4.9   There would be 33 H bonds within the helix. Since an α-helix contains 3.6 amino acid residues per turn, there are 36/3.6 or 10 turns within this helix.

The number of H bonds is best determined by studying Figure 4.3. Since each internal amino acid residue contributes to two H bonds, it is tempting to multiply the total number of amino acids by two, and then subtract 2 for the two terminal residues. In doing so, however, you would be counting each H bond twice; the H bond to the amide carbonyl group in one amino acids residue is the same as the H bond to an amide hydrogen further along the chain. Counting the number of amino acid residues and then subtracting 3 yields the correct number of H bonds. One must subtract 3 because the amide hydrogens in the last 3 residues at the N-terminus of an α-helix have no amide carbonyls to bond with further along the helix (there is no further helix). Each of the other amide hydrogens participate in the formation of 1 H bond. Five H bonds hold the 8 residues together within the α-helix shown in Figure 4.3 (be careful and do not count H bonds to residues not shown).

4.10

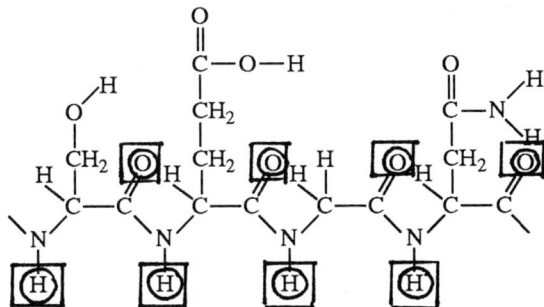

A study of Figures 4.3 and 4.5 indicates that both sheets and helices are maintained through H bonding between amide hydrogens and amide carbonyls. The fragment would most likely be located on the surface of a globular protein since 3 of the 4 residues have polar side chains—nonpolar in, polar out (the general rule for the folding of globular proteins in an aqueous environment).

4.11   Yes. In theory, virtually any coiling pattern is possible. In practice, the most stable coiling patterns are usually the ones described in this chapter. Helix stability is largely determined by the number and strength of the H bonds involved in maintaining the helix. Other factors, such as steric hindrance and repulsions between like charges, can also impact relative stability. The α-helix is the most common helix found within proteins.

4.12

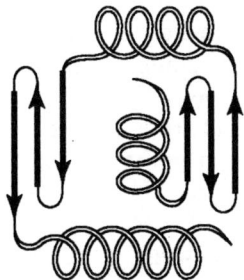

There are many possible answers.

4.13   Primary structure—peptide (amide) bonds

Secondary structure—H bonds

Tertiary structure—disulfide bonds; H bonds; salt bridges; hydrophobic forces; van der Waals forces

Quaternary structure—H bonds; salt bridges; hydrophobic forces; van der Waals forces

The covalent peptide bonds and disulfide bonds are the strongest.

4.14   The π-helix has nine complete residues plus a partial residue at each end. The 3_{10}-helix contains five complete residues plus a partial residue at each end. Quickly count the α carbons or R groups (some partially or totally hidden), and then carefully examine the terminal residues.

4.15   pH 1: Arg, Lys and His; pH 7: Arg, Asp, Glu and Lys; pH 12: Asp, Glu, Cys and Tyr.

Since a side chain must have a net charge to participate in salt bridge formation, this question could be reworded: "which amino acids have a side chain with close to a unit net charge at the pH's of interest?" To answer this question, one must compare each pH to the pK_a's of the weak acid side chains (Table 3.5). If the pH is 0.5 or more below the pK_a, a side chain is assigned the charge of its weak acid form for charge calculations to the nearest half of a unit. At pH's 0.5 or more above the pK_a, a weak acid functional group is assumed to exist entirely in its conjugate base form for charge calculations to the nearest half of a unit. In every case, either the weak acid or the conjugate base of a functional group will be charged. For some functional groups (not present in amino acids), both the weak acid and conjugate base are charged.

4.16   No.

To be part of α-helix, an amino acid residue must H bond to other residues within the helix and have unique values for ϕ and ψ. Similarly, an amino acid residue in a sheet must H bond to other residues in the sheet and must have characteristic values for ϕ and ψ that are incompatible with an α-helix. Once H bonded within a sheet, a residue would be unable to H bond appropriately within a helix and vice versa; the same atoms are involved in H bonding in both instances. This can best be appreciated by building or studying a model.

4.17   When folded into the tertiary structure of a protein, one side of a helix may find itself interacting with the nonpolar core while the opposite side is on the surface interacting with water: nonpolar in, polar out.

4.18   Internal (mostly nonpolar): (Gly)(Ala)Leu-Pro-Trp-Phe-Ser-Met(Cys)(Gly) and (Tyr)(Ala)Ile-Phe-Met-Pro-Cys-Val

External (predominantly polar): Arg-Cys-Pro-Gly-Ser-Glu-Asn(Gly)(Ala) and (Cys)(Gly)His-Tyr-Glu-Asn-Thr-Lys(Tyr)(Ala)

The answer is based on the general folding rule (nonpolar in, polar out) plus Table 4.5. For the variable residues in parentheses, it is impossible to predict whether they would be part of internal or external segments.

4.19   In an α-helix (3.6 amino acid residues per turn) with one side facing in and one side out, every third or fourth residue would tend to be on the same side of the helix and of similar polarity (polar if facing out and nonpolar if facing in; see Exercise 4.17).

4.20   Extracellular proteins tend to be subjected to harsher environments and a greater variety of insults than intracellular ones. The strong, covalent disulfide bonds are needed to protect the extracellular proteins from denaturation. In general, proteins with disulfide bonds are more difficult to unfold than those without.

4.21   Two helix-turn-helix motifs are illustrated in Figure 4.11. A β-barrel and α/β-barrel are depicted in Figure 4.12. These are the motifs examined in greatest detail within this chapter. Although they are described in the text, an original description should be possible based on a study of Figures 4.11 and 4.12. The purpose of this question is to encourage you to study these motifs and to think a bit about their structural features.

4.22   Heat, extremes in pH, organic solvents, high salt concentrations, mechanical agitation, and detergents are all denaturing "agents" (see Exhibit 4.1).

Quaternary structure tends to be denatured (disrupted through unfolding) by any treatment that ruptures the relatively weak noncovalent bonds (H bonds, hydrophobic forces, salt bridges, and van der Waals forces) that maintain this structural element. Tertiary and secondary structures are denatured by similar treatments.

4.23   Primary structure: amino acid sequence.

Secondary structure: the arrangement of segments of primary structure into helixes and β-sheets.

Motifs: recurring patterns in which secondary structural elements are folded together.

Domains: independently folded segments along a polypeptide chain. May contain multiple motifs.

Tertiary structure: the overall three-dimensional folding of a single polypeptide chain; the folding of individual secondary structural elements, motifs, "random regions," and domains with respect to one another.

Quaternary structure: the packing together of separate polypeptide chains in those proteins that contain multiple, noncovalently-bonded polypeptide chains. In a globular protein, each polypeptide chain is folded into a tertiary structure.

4.24   Chaperones act by blocking off-pathway folding and by providing an environment that favors on-pathway folding. Chaperones are, themselves, proteins.

4.25   The shift in pH leads to the dissociation of the subunits within an oligomer. Because they are smaller than the oligomer, the subunits more readily enter the pores of gel filtration beads. Consequently, they progress more slowly than the oligomer as they work their way through the column. A single peak at pH 1 indicates that all of the subunits must be of similar size.

4.26   Since a nutritionally-complete protein contains all of the essential amino acids (Table 3.7), the primary structure of myoglobin must encompass all of these amino acids.

4.27   Ethanol tends to kill microorganisms by inactivating their proteins.

Ethanol is a rather unique solvent; it has a reasonably good balance between polar and nonpolar character. The nonpolar part of its character tends to turn a globular protein inside out or at least weaken the hydrophobic forces that normally play a major role in the maintenance of its native structure. In a nonpolar environment, the folding rule is "polar in, nonpolar out" rather than "nonpolar in, polar out".

4.28   8,066.

A single water molecule has a volume of about 6×10^{-3} cubic nm ($0.1 \times 0.2 \times 0.3$) while myoglobin occupies approximately 48.4 cubic nm ($4.4 \times 4.4 \times 2.5$). Dividing 48.4 by 6×10^{-3} yields 8,066. A single myoglobin molecule is over 8,000 times larger than a water molecule.

4.29   Virtually the entire surface of a myoglobin molecule is coated with polar or charged side chains that would rather interact with surrounding water molecules (very polar) than neighboring myoglobins. The subunits of hemoglobin clump together because each subunit has a nonpolar side that wants to escape from the polar aqueous environment.

4.30   Probably not. One of the histidines interacts directly with the O_2 in oxymyoglobin and plays a crucial, direct role in its binding. The other histidines helps hold the heme group in the heme pocket and helps position the heme in such a manner that it is able to bind O_2. Figure 4.19 documents these claims. The small nonpolar Ala side chain could not replicate the functioning of the larger, highly polar histidine side chain.

4.31   Most of the bends and loops are located on the surface of myoglobin rather than being buried inside. Surface residues tend to be polar since myoglobin is normally surrounded by water—"nonpolar in, polar out". Since at least one side of each helix is part of the nonpolar core, it is coated mainly with nonpolar side chains.

4.32   Since many life sustaining processes, including the synthesis of proteins and most other classes of compounds, use energy, each cell in the body requires a virtually constant supply of energy to survive. Most of this energy is produced during the oxidation of dietary fuels (organic matter $+ O_2 \rightarrow CO_2 + H_2O +$ energy). Since O_2 is required for this oxidation, when a cell runs out of O_2, it runs out of energy and malfunctions or dies.

4.33   Pairs 1, 4 and 5.

Two compounds must differ significantly in size or shape before they can be separated readily from one another with gel filtration chromatography. Hemoglobin is about four times the size of myoglobin, and myoglobin, with over 150 amino acids, is many times larger than either alanine or a decapeptide. HbA and HbF have exactly the same number of amino acid residues and would be predicted to have virtually the same size and shape. The α chain of hemoglobin has 141 amino acids and the β chain 146. Differing by less than 4% in number of residues, these chains would be difficult to separate.

In theory, molecular weight, compactness, and shape will all have some impact on how readily a molecule can enter the pores in gel filtration beads. In practice, most organic molecules of similar molecular weight are similar enough in overall size and shape that they travel at about the same rate through a gel filtration column.

4.34   Less tightly. Since both the heme group and heme pocket are predominantly nonpolar in nature, a charged, polar group in the pocket would disrupt and destabilize the hydrophobic interactions principally responsible for holding the heme group in its pocket.

4.35   A decrease in the K_{eq} for O_2 binding will cause the O_2 binding curve to fall to the right of the original curve. With a smaller equilibrium constant, it will take a higher concentration of O_2 to oxygenate any given fraction of the hemoglobin molecule. Viewed from a slightly different perspective, the amount of oxyhemoglobin present at equilibrium will decrease as the equilibrium constant drops (assuming that all other reaction conditions remain constant).

4.36   The cooperative binding of O_2 by hemoglobin is a consequence of hemoglobin's quaternary structure. In normal hemoglobin, the binding of one O_2 changes the conformation of that subunit to which it binds. This change in conformation modifies the interactions between subunits, and, in the process, alters the conformations of those subunits not yet bound to O_2. Thus, the subunits communicate with one another through quaternary structure interactions. In an all β tetramer, the communication between subunits has been modified to such an extent that the binding of O_2 to one polypeptide has no impact on the oxygen binding ability of neighboring polypeptides within the tetramer.

4.37   Nonpolar in, polar out. The larger a globular protein, the larger the fraction of residues inside (nonpolar ones) and the smaller the fraction on the surface (polar ones). The surface area of a sphere is proportional to the square of the radius while the volume is proportional to the cube of the radius.

4.38   Myoglobin has one domain, hemoglobin four. A domain is an independently folded region along a single polypeptide chain; if the rest of the chain is "cut" away, the folding will be unmodified. Within a globular protein, every polypeptide chain has at least one domain. Each of the polypeptide chains in myoglobin and hemoglobin contains a folded region which encompasses the entire chain.

4.39   Since a majority of the CO_2 reacts with H_2O to form H_2CO_3 which then dissociates into H^+ and HCO_3^-, most of the CO_2 is transported as HCO_3^-. Some of the CO_2 reacts with the N-terminal α-amino groups on the α and β chains of hemoglobin to form carbamates. This CO_2 is carried to the lungs covalently bound to hemoglobin. A small fraction of the CO_2 is transported as a nonreacted, dissolved gas. CO_2 transport is summarized in Table 4.6.

4.40   Acute CO poisoning can be treated by increasing lung O_2 concentration. Since both O_2 and CO bind reversibly to the same site, they compete with one another for that site. Consider the following reactions:

$$Hb + 4CO \rightleftarrows Hb(CO)_4$$

$$Hb + 4O_2 \rightleftarrows Hb(O_2)_4$$

If more O_2 is added to a system at equilibrium, the second reaction will shift to the right. In the process, the concentration of free Hb drops and causes the first reaction to shift to the left (LeChâtelier's Principle). The net result? Oxygen has taken Hb away from the CO.

4.41   In the presence of CO, a higher O_2 concentration is required to attain any specified level of O_2 saturation. Consequently, the O_2 binding curve falls to the right of the normal curve. CO and O_2 compete for the same binding sites on hemoglobin.

4.42   Because BPG is an anion, it cannot pass through the nonpolar layer within the membranes surrounding red blood cells. Unless specific carriers or channels exist, polar and charged substances usually find membranes impermeable (Section 9.10). Polar and nonpolar do not mix well.

4.43   Less. At extremely high red blood cell BPG concentrations, the binding curve will fall so far to the right that hemoglobin leaving the lungs will carry very little O_2. Consequently, it will be unable to deliver a significant amount of O_2 to tissues:

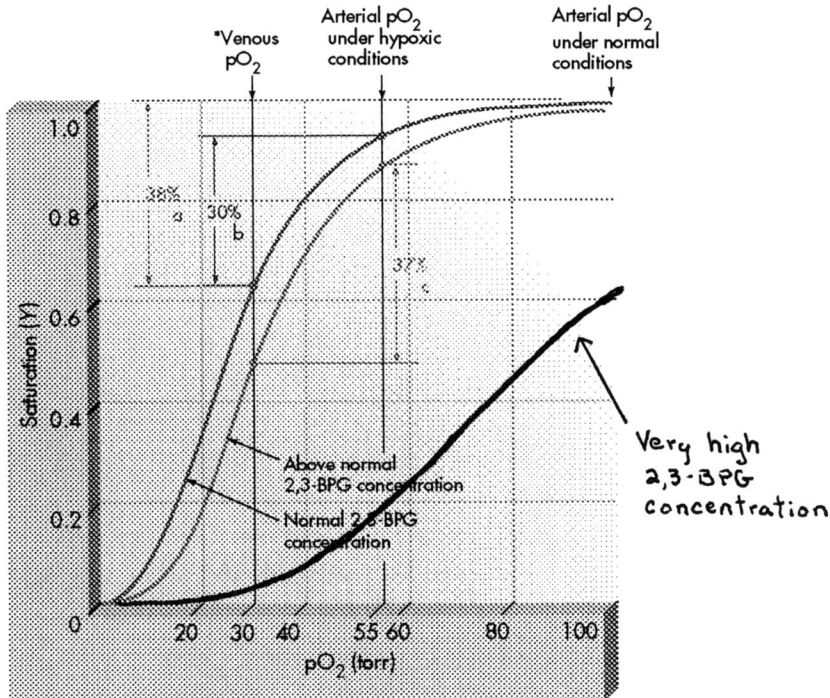

*Venous pO_2 = pO_2 in blood returning to the lungs.
Arterial pO_2 = pO_2 in blood leaving the lungs.

4.44

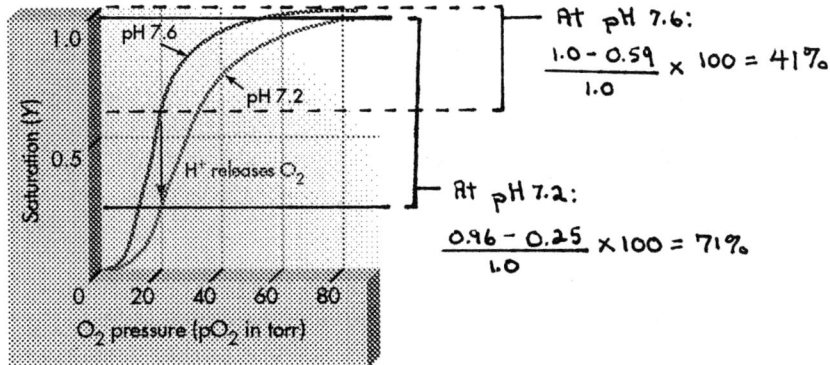

At pH 7.6:

$$\frac{1.0 - 0.59}{1.0} \times 100 = 41\%$$

At pH 7.2:

$$\frac{0.96 - 0.25}{1.0} \times 100 = 71\%$$

4.45   At very high pH's, hemoglobin, like most other globular proteins, is denatured. When denatured, hemoglobin releases its heme groups and is unable to bind any O_2.

4.46   The toughness of steak from an old steer is mainly a consequence of the age-linked accumulation of cross-linked macromolecules, principally collagen. Covalent cross-linking tends to make molecules physically tough.

4.47

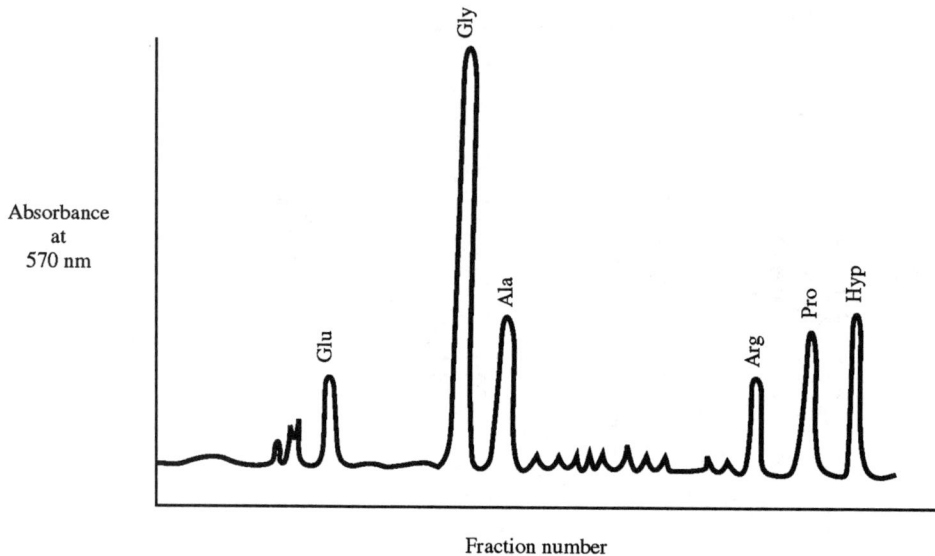

Fraction number

The actual location of each amino acid will vary with chromatographic conditions. The important concept is that the peaks for Gly, Pro, Hyp, Ala, Glu, and Arg will be much higher than the peaks for the other amino acids, since these are the predominant amino acids in collagen. Gly, alone, accounts for roughly 33% of the residues in collagen. Pro plus Hyp account for another 21%.

4.48   Hydroxyproline and hydroxylysine. These unusual residues are produced through the enzymatic modification of proline and lysine residues after they have been incorporated into a collagen polypeptide. Unless the hydroxylations occur after a residue has been incorporated into a growing polypeptide but before polypeptide assembly has been completed, a nascent (newly synthesized) collagen polypeptide will be free of Hyp and Hyl residues.

4.49   The catalysts that digest proteins (catalyze the hydrolysis their peptide bonds) have a difficult time accessing the peptide bonds within the twisted and covalently cross-linked polypeptide strands in hair and collagen. While globular proteins tend to be denatured within the digestive system (exposes their peptide bonds), the covalent cross-links in hair and collagen make them resistant to unraveling.

4.50   Low quality. Collagen is deficient in some of the essential amino acids.

4.51   12 total domains: 2 within each light chain and 4 within each heavy chain. Since all domains contain β-sheets but no helices, the immunoglobulin folds are classified as β-domains.

4.52   Yes. Cleavage at the hinge would have no impact on the conformation of the antigen binding site. Once bound, however, the fragment would be unable to orchestrate a normal immune response against the antigen; an intact IgG is required for this process. It is principally the base of the "Y" which communicates with the rest of the immune system.

4.53   Primary structure: the sequence of amino acids within each of four polypeptide chains.

Secondary structure: the multiple antiparallel β-pleated sheets formed along each chain.

Tertiary structure: the folding of each chain into multiple domains with each domain containing two β-sheets stacked one on top of another.

Quaternary structure: None. One does not normally talk about quaternary structure unless multiple polypeptide chains are held together through noncovalent bonds. The four chains in IgG are linked through covalent disulfide bonds.

4.54   NMR spectroscopy can be performed on proteins in solution under what approximates physiological conditions. Solid protein crystals are required for X-ray analysis.

At the present time, only relatively small proteins can be analyzed with NMR techniques. There is no limit on the size of a protein that can be examined with X-ray diffraction.

4.55   The number of peaks corresponds to the number of chemically distinct classes of carbons.

a) 2 peaks: there are two chemically distinct classes of carbons; the carbonyl carbon is in a class by itself while the two methyl carbons are equivalent.

b) 3 peaks: each of the three carbons is in a unique environment.

c) 5 peaks: each carbon is in a unique environment.

5

Enzymes

5.1

Reaction A

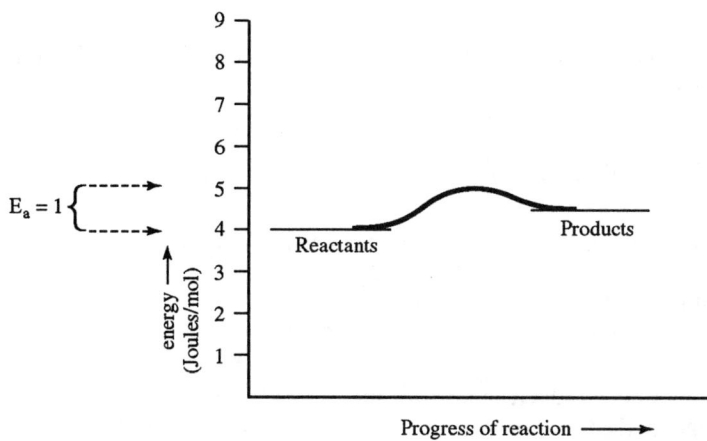

$E_a = 1$

energy (Joules/mol)

Reactants

Products

Progress of reaction ⟶

With a low activation energy, this reaction will tend to occur relatively rapidly.

Reaction B

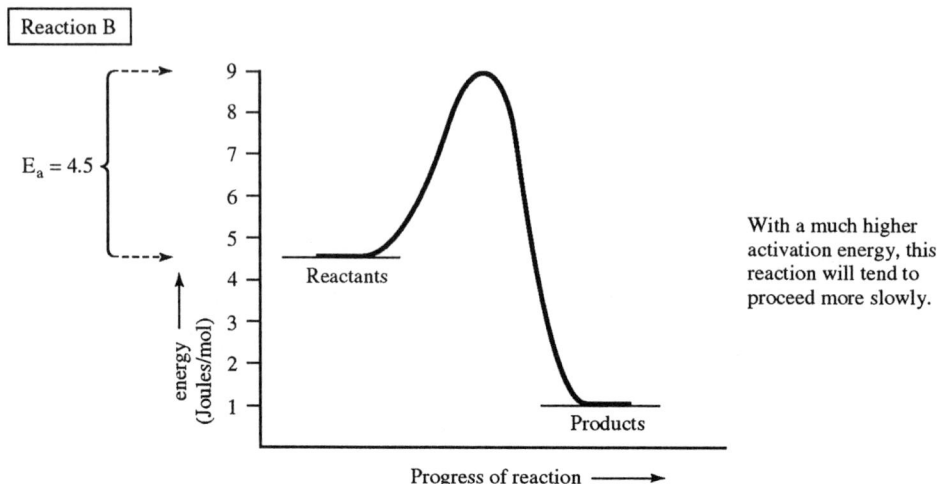

$E_a = 4.5$

energy (Joules/mol)

Reactants

Products

Progress of reaction ⟶

With a much higher activation energy, this reaction will tend to proceed more slowly.

All other factors (nature of reactants, collision frequency, temperature, etc.) being approximately equal, the reaction with the lowest activation energy will tend to occur most rapidly.

Since Reaction B has the largest activation energy, it would benefit the most from an effective catalyst, one that lowers the activation energy to near zero.

5.2   Since one E yields two P, the rate of disappearance of E is 2×10^{-3} mmol/10 s = 2×10^{-4} mmol/s.

$$\frac{2 \times 10^{-4} \text{ mmol}}{1 \text{ s}} \times \frac{60 \text{ s}}{1 \text{ min}} = 0.012 \text{ mmol/min.}$$

P is appearing at a rate of 4×10^{-3} mmol/10 s = 4×10^{-4} mmol/s.

$$\frac{4 \times 10^{-4} \text{ mmol}}{1 \text{ s}} \times \frac{60 \text{ s}}{1 \text{ min}} = 0.024 \text{ mmol/min.}$$

With 4×10^{-3} mmol of product, the rate of the reverse reaction is probably insignificant relative to the rate of the forward reaction. If this is the case, the calculated rates are a good measure of the initial velocity for the reaction. Initial velocity, by definition, is the rate before the reverse reaction must be taken into consideration.

5.3   An international unit (U), by definition, is that amount of enzyme that converts 1 μmol of substrate to product per minute under standard reaction conditions. Since rate is proportional to enzyme concentration when substrate is saturating (Figure 5.6), a rate of 0.02 μmol/min corresponds to 0.02 U of enzyme.

$$0.02 \text{ U} \times \frac{1.67 \times 10^{-8} \text{ kat}}{1 \text{ U}} = 3.34 \times 10^{-10} \text{ kat}$$

To calculate turnover number, one needs the total number of enzyme molecules responsible for the observed rate. This can be calculated given the grams of enzyme present and the molecular weight of the enzyme.

5.4   Same number. Under standard conditions and in the absence of activators or inhibitors, a molecule of enzyme will have the same catalytic activity, regardless of what other molecules are present along with it. Since U is defined in terms of catalytic activity, one U of a given enzyme will always contain the same number of enzyme molecules regardless of the purity of the enzyme preparation (assuming that no activators or inhibitors are present).

5.5   The specific activity of the crude extract is 5.0×10^8 U/mg:

$$\frac{5.0 \times 10^{12} \text{ U}}{10 \text{ g}} \times \frac{1 \text{ g}}{1000 \text{ mg}} = 5.0 \times 10^8 \text{ U/mg}$$

The specific activity after chromatography is 1.7×10^9 U/mg:

$$\frac{1.2 \times 10^{12} \text{ U}}{0.7 \text{ g}} \times \frac{1 \text{ g}}{1000 \text{ mg}} = 1.7 \times 10^9 \text{ U/mg}$$

A 3.4-fold purification was attained with the chromatography:

$$\frac{1.7 \times 10^9}{5.0 \times 10^8} = 3.4$$

5.6   Absolutely nothing. Rates are primarily determined by collision frequencies, activation energies and catalysts while K_{eq}'s are, at a constant temperature, determined by the free energy difference between reactants and products. The K_{eq} for a reaction provides no clues as to the rate of that reaction and vice versa.

5.7   Bimolecular. Water is one of the two reactants. Hydrolysis, by definition, is the splitting of a compound into two parts through reaction with water. The water contributes a H to one product and an OH to the second.

5.8   A rise in temperature increases the K_{eq} for an endothermic reaction while decreasing the K_{eq} for exothermic reactions. This fact is predictable from Le Châtelier's principle if one treats heat as a reactant or product.

5.9   All enzymes are, by definition, catalysts. However, not all catalysts are enzymes. The catalytic converters on automobiles contain heat-resistant inorganic catalysts, not enzymes. A wide variety of non-enzyme catalysts, including acids, bases and platinum, are utilized by organic chemists for synthetic purposes. A catalyst must be of biological origin to be classified as an enzyme. All known enzymes are either proteins or RNAs.

5.10   Enzymes are such potent catalysts that relatively small amounts of an enzyme can normally satisfy the biological need for that protein. In contrast, many other proteins must be assembled in much larger quantities in order to fulfill their biological roles or functions.

5.11   Short. To respond to rapidly changing needs and activities, a cell must be able to rapidly change the combination of reactions occurring within it. In general, each reaction requires a separate protein catalyst (enzyme). Many of the enzymes needed at one instance may differ from those required 5 minutes later. In fact, some of the reactions essential at one instance may be detrimental if they continue to occur 5 minutes later. Enzymes tend to be synthesized when they are needed and to be degraded when they are no longer required. This avoids storage problems and helps insure that only appropriate reactions are occurring at any given time.

5.12   Free amino acids and some ammonia (if the enzyme contains one or more glutamine or asparagine residues). The only hydrolyzable bonds in a simple enzyme are the peptide bonds and the amide links within glutamine and asparagine side chains. The hydrolysis of complex enzymes (contain nonpeptide prosthetic groups) will yield additional products as well.

5.13   Prosthetic groups are nonpolypeptide components within proteins. Prosthetic groups within enzymes are called cofactors. Organic cofactors are classified as coenzymes.

5.14   The binding site on an enzyme is chiral. A baseball glove (they are all chiral) provides a good analogy: the left hand (also chiral) will not fit into a right-handed glove. Similarly, the binding site for an L-enantiomer will not accommodate a D-enantiomer.

5.15   "G," none of the above, should be circled. According to the lock and key model, an enzyme has a fixed, predetermined structure and shape that does not change when the substrate binds. The act of binding does not chemically alter the enzyme.

5.16   "C" should definitely be circled and "b" in some instances. According to the induced fit model, an enzyme is induced to fit around its substrate as that substrate binds. Since an enzyme must change its conformation to modify its fit, the tertiary structure (a central component of conformation) of the protein must be altered. Secondary structure may be modified as well. It is likely that quaternary structure, if present, would also be impacted. Once again, we would not expect substrate binding to lead to hydrolysis, denaturation or a change in pH optimum.

5.17

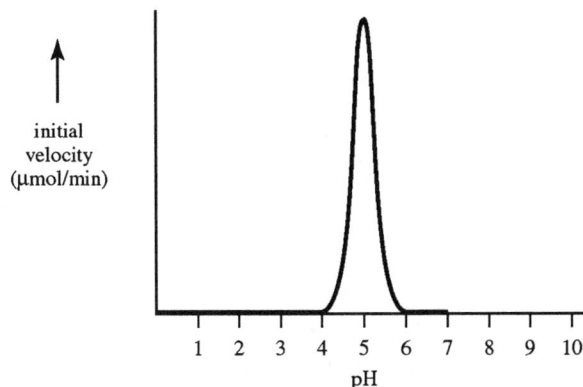

5.18   The shift in pH will lead to the partial or total denaturation of numerous proteins, including many enzymes. This denaturation will impact the functioning of most of these proteins, and, in the process, disrupt some of the reactions and processes essential to life.

5.19   This relationship minimizes the number of enzyme molecules that a cell must assemble to attain any given reaction rate. Both energy (required to synthesize enzymes) and space (occupied by enzymes) are conserved in the process.

5.20   No. The sensitivity to pH changes depends on multiple factors including initial pH, the pK_a's of the weak acid functional groups, the role of these functional groups in the binding and catalytic processes, and their role in the maintenance of a native conformation.

5.21   D. Glutamic acid. Like charges repel, and both oxalate and glutamic acid are negatively charged at pH 7.4. Positively-charged binding sites tend to have a high affinity for negatively-charged substrates.

5.22   The enzyme has a very similar conformation and very similar properties at both pH 6.5 and 7.6. The substrates also have similar properties and structures at pH 6.5 and 7.6. In contrast, the nature of the inhibitor (net charge, shape, etc.) is significantly different at these two pH's.

5.23   Remain the same. At saturating substrate concentrations, all of the enzyme is bound to substrate and in the process of converting substrate to product. Since the enzyme is operating at full capacity, the initial velocity is already maximized and the further addition of substrate leads to no increase in reaction rate.

5.24   Increase. Le Châtelier's principle predicts that the addition of more reactant to a reaction at equilibrium will lead to a net forward reaction until equilibrium is restored. The addition of the reactant temporarily destroys the equilibrium. The enzyme is a reactant during the formation of the enzyme-substrate complex:

$$E + S \rightleftarrows ES$$

This explains why reaction rates increase as enzyme concentrations increase (all other reaction conditions held constant).

5.25   Eighteen percent of 25 or 4.5 mmol/min.

Work accomplished is directly proportional to the percentage of the workforce that is actually working. At V_{max}, all (from a practical standpoint) of the enzyme is bound to substrate and working all of the time. If half of the enzyme is bound to substrate, the rate is one half V_{max}. When 18% of the enzyme is working, the rate is 18% of V_{max}.

5.26 K_m's provide no information about V_{max}'s. K_m's are determined principally by the binding sites on enzymes while V_{max}'s are governed primarily by catalytic sites. V_{max} is the initial velocity when all of the enzyme is working all of the time. It is independent of the substrate concentration required to saturate the enzyme.

In a reaction mixture that contains equal concentrations of both Z and W, the EW complex will be present in much higher concentrations than the EZ complex. Although not always true, you have been asked to assume that the K_m is at least a rough measure of the dissociation constant for the enzyme-substrate complex. The smaller this dissociation constant, the more tightly an enzyme binds a substrate and the greater the concentration of the enzyme-substrate complex at equilibrium. Since compound W has a much smaller K_m than compound Z, the enzyme has a considerably greater affinity for compound W than compound Z.

5.27   Very small. A small K_m means that the enzyme has a high affinity for the toxin (see answer to Exercise 5.26) and is able to effectively bind the toxin, even at low toxin concentrations. An enzyme with a very large K_m would be unable to effectively bind the toxin until the toxin had already reached lethal levels. An enzyme cannot modify a substrate until it has bound that substrate, regardless of how potent its catalytic site might be.

5.28   Binding sites are highly specific and very discriminating. Even minor changes in a substrate can impact its interactions with the amino acid side chains that constitute the binding site. The modified interactions normally lead to altered binding constants and K_m's.

5.29   True. K_m, by definition, equals the substrate concentration that leads to an initial velocity equal to half V_{max} (attained when the enzyme is half saturated). The substrate concentration that half saturates the enzyme is the same regardless of the substrate concentration that actually exists. Similarly, the maximum number of basketballs you are capable of carrying is unaffected by the number of basketballs you are carrying at any instance.

5.30   No. Equilibrium constants are independent of the concentrations of reactants and products. Changes in reactant or product concentrations may temporarily destroy an equilibrium, but they do not change the equilibrium constant itself.

5.31   Decrease. The enzyme will be operating further down its v_o versus [substrate] curve; at any given substrate concentration, the fraction of enzyme bound to substrate decreases as K_m increases. Restated, an increase in K_m indicates that the enzyme's affinity for its substrate has decreased.

5.32   5.79 μmol/s.

$$v_o = \frac{V_{max}[S]}{[S] + K_m} = \frac{(22.2)(0.00012)}{(0.00012) + (0.00034)} = 5.79 \text{ μmol/s.}$$

5.33

v_o (mmol/s)	[S] (M)	$\frac{1}{v_o}$ (s/mmol)	$\frac{1}{[S]}$ M^{-1}
0.024	1.0×10^{-4}	42	1×10^4
0.031	1.7×10^{-4}	32	5.9×10^3
0.036	2.5×10^{-4}	28	4×10^3
0.045	5.0×10^{-4}	22	2×10^3

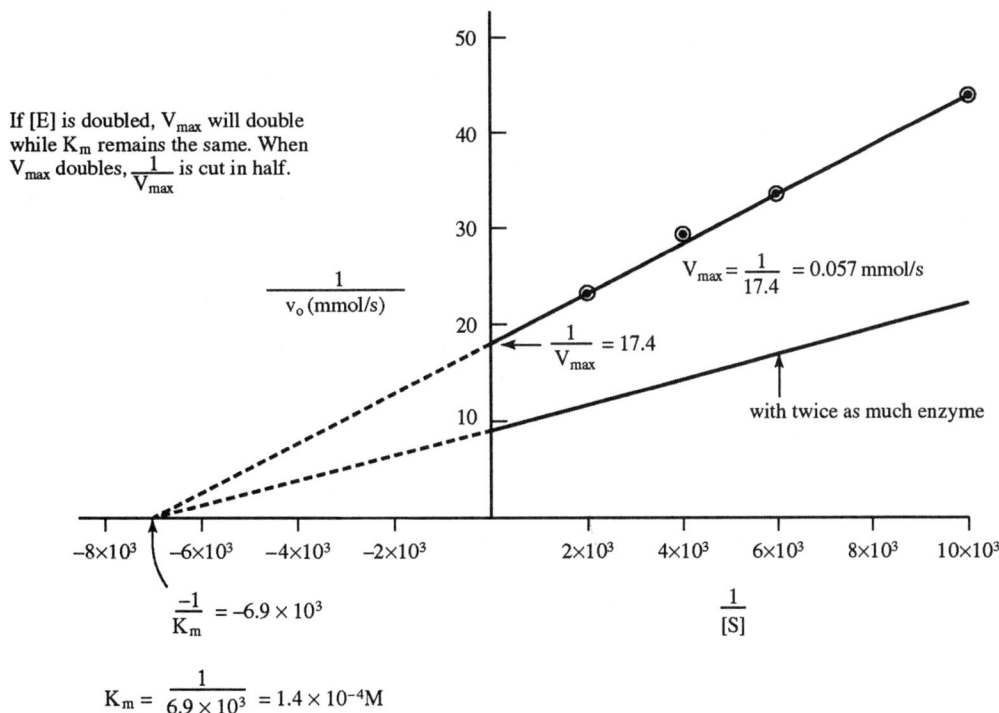

$1 \times 10^4 \longleftarrow = 10 \times 10^3$

If [E] is doubled, V_{max} will double while K_m remains the same. When V_{max} doubles, $\frac{1}{V_{max}}$ is cut in half.

$\frac{1}{v_o}$ (mmol/s)

$V_{max} = \frac{1}{17.4} = 0.057 \text{ mmol/s}$

$\frac{1}{V_{max}} = 17.4$

with twice as much enzyme

-8×10^3 -6×10^3 -4×10^3 -2×10^3 2×10^3 4×10^3 6×10^3 8×10^3 10×10^3

$\frac{-1}{K_m} = -6.9 \times 10^3$

$\frac{1}{[S]}$

$K_m = \frac{1}{6.9 \times 10^3} = 1.4 \times 10^{-4} M$

5.34

[S] (M)	v_o with no inhibitor (mmol/min)	v_o with inhibitor (mmol/min)	$\frac{1}{[S]}$ M^{-1}	$\frac{1}{v_o}$ with no inhibitor (min/mmol)	$\frac{1}{v_o}$ with inhibitor (min/mmol)
2.0×10^{-3}	0.123	0.057	500	8.1	17.5
2.5×10^{-3}	0.133	0.067	400	7.5	14.9
3.3×10^{-3}	0.145	0.081	303	6.9	12.3
5.0×10^{-3}	0.161	0.100	200	6.2	10.0
10.0×10^{-3}	0.175	0.135	100	5.7	7.4

Inhibitor is competitive since, in the presence of inhibitor, K_m increases while V_{max} remains the same. The K_m values on this diagram are "apparent K_m" values (the substrate concentration that leads to an initial velocity equal to one half V_{max}).

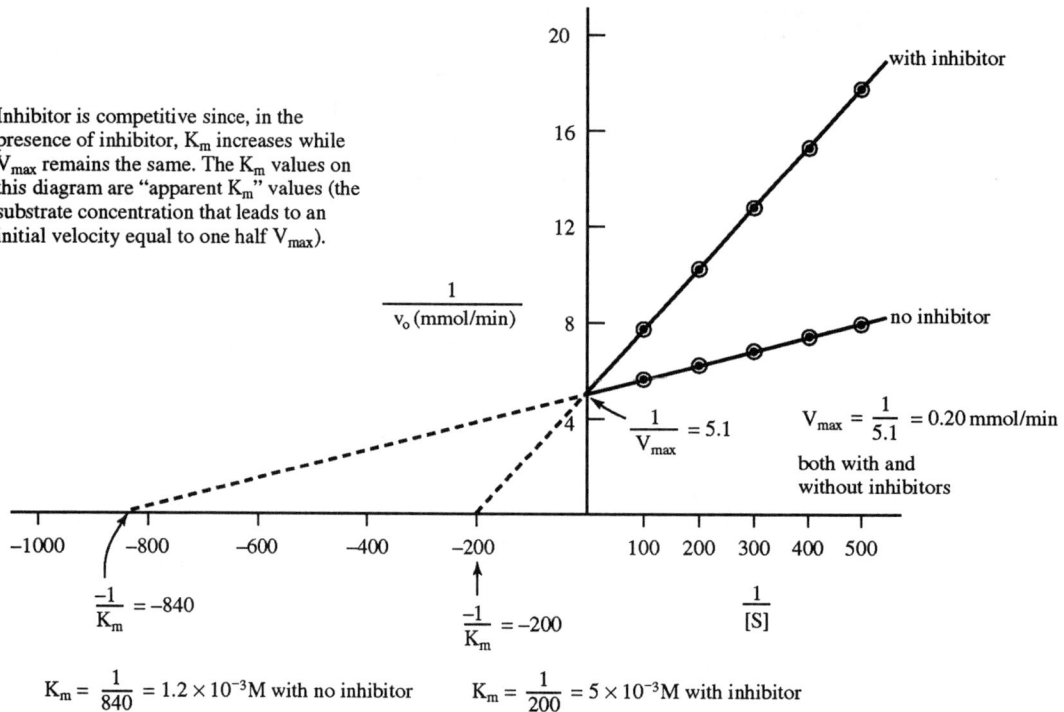

$\frac{1}{V_{max}} = 5.1$

$V_{max} = \dfrac{1}{5.1} = 0.20 \, \text{mmol/min}$

both with and without inhibitors

$\dfrac{-1}{K_m} = -840$

$\dfrac{-1}{K_m} = -200$

$K_m = \dfrac{1}{840} = 1.2 \times 10^{-3} \text{M}$ with no inhibitor $K_m = \dfrac{1}{200} = 5 \times 10^{-3} \text{M}$ with inhibitor

5.35 $E + S \underset{k_{-1}}{\overset{k_1}{\rightleftarrows}} ES \overset{k_2}{\rightarrow} E + P$

5.36 $E + P \underset{k_2}{\overset{k_{-2}}{\rightleftarrows}} ES$

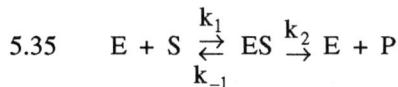

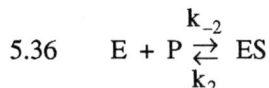

Rate of forward reaction $= k_{-2}[E][P]$

Rate of reverse reaction $= k_2[ES]$

At equilibrium, $k_{-2}[E][P] = k_2[ES]$. Dividing both sides of this equation by $[E][P]$ and k_2:

$$\frac{k_{-2}}{k_2} = \frac{[ES]}{[E][P]} = K_{eq} \text{ for reaction}$$

5.37   Many enzymes have never been completely purified and accurate molecular weights are available for a limited number of enzymes. Even if you know the number of grams of a specific enzyme in a sample, you cannot convert grams to moles without molecular weight information.

 When working with a sample of an enzyme, what is usually most important is the catalytic activity of that sample, not its weight or the number of moles of enzyme present. Since enzymes are very sensitive critters, only 80% of the enzyme molecules in a sample of a purified enzyme may be active. The rest may have been denatured or inactivated in some other manner. In such a case, grams or moles would provide no information about the quantity of functional enzyme within the sample.

5.38   Turnover number is defined as the number of substrate molecules converted to product per unit time (minute, second, etc.) with the catalytic assistance of a single molecule of enzyme. To make such a calculation, one must know the total number of enzyme molecules in the reaction mixture where the rate of the reaction is being measured. The total number of enzyme molecules can only be determined if one has added (to the reaction mixture) a known mass of an enzyme of known molecular weight. The known mass (expressed in grams) and the molecular weight allow one to calculate the moles of enzyme:

$$\text{mass in grams} \times \frac{\text{one mole}}{\text{molecular weight (given units of grams)}} = \text{moles}$$

One mole of enzyme contains 6.02×10^{23} (Avogadro's number) enzyme molecules.

5.39   The K_m for ester A will increase in the present of ester B. Ester B is a competitive inhibitor of the hydrolysis of ester A, and vice versa, since the two compounds compete for binding to the same active site (an example of substrate inhibition). In the presence of a competitive inhibitor, it takes more substrate to half saturate an enzyme with substrate. Consider the following two reactions:

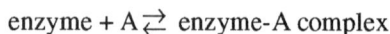

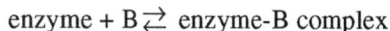

enzyme + A $\rightleftarrows$ enzyme-A complex

enzyme + B $\rightleftarrows$ enzyme-B complex

Think Le Châtelier's principle. When the addition of B pulls some enzyme into an enzyme-B complex, it takes more A to pull half of the total enzyme into an enzyme-A complex. Ester A must compete the enzyme away from ester B.

The V_{max} for the hydrolysis of ester A will remain the same. At high enough concentrations of A, the first reaction (above) will shift completely to the right and, in the process, pull the second reaction completely to the left. All of the enzyme (from a practical standpoint) will be bound to A and in the process of catalyzing the hydrolysis of this substrate.

5.40   The reversible binding of the inhibitor changes the conformation of the enzyme to such an extent that the substrate can no longer bind. Similarly, when the substrate binds, it alters the conformation of the enzyme such that inhibitor is unable to bind. The two substances must compete for the enzyme since they cannot both bind to the enzyme at the same point in time.

5.41   a) I′ is the most potent inhibitor. The equilibrium constant for the binding of I′ to enzyme W is 2000 times larger than the equilibrium constant for I binding.

b) EI′. Once again the answer is based on the value of the reported equilibrium constants. The binding constant for I′ is larger than the binding constant for either I or substrate.

c)

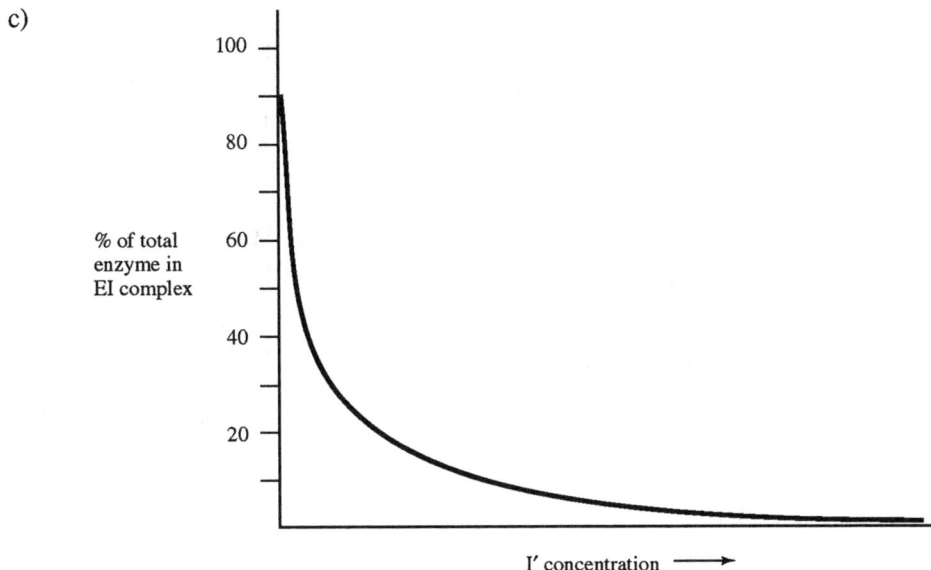

% of total
enzyme in
EI complex

I' concentration ⟶

One obtains an inverted hyperbolic curve. I' competes with I for the active site, and at very high I' concentrations all of the I is displaced from the enzyme.

Since the binding constant for I' is comparatively large, it takes relative little I' to pull most of the free enzyme into an EI' complex. As free enzyme concentration drops, I is released from the EI complex (Le Châtelier's principle) and more free enzyme is generated. Thus, the concentration of free I increases as I' is added. If still more I' is added, the reaction of E with I' shifts further towards the EI' complex and more I is released from the EI complex. At extremely high I' concentrations, virtually all of the enzyme will exist in an EI' complex and there will be no significant amount of either free enzyme or EI complex; all of the I will exist in a free, nonbound form.

5.42 a)

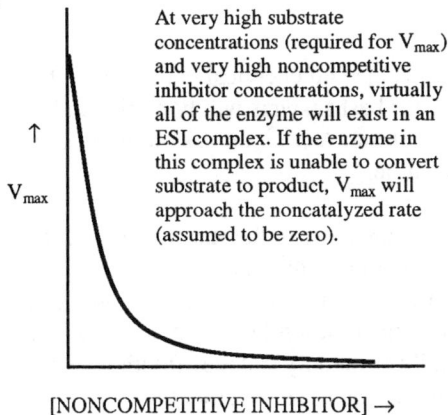

↑

V_{max}

At very high substrate concentrations (required for V_{max}) and very high noncompetitive inhibitor concentrations, virtually all of the enzyme will exist in an ESI complex. If the enzyme in this complex is unable to convert substrate to product, V_{max} will approach the noncatalyzed rate (assumed to be zero).

[NONCOMPETITIVE INHIBITOR] →

b)

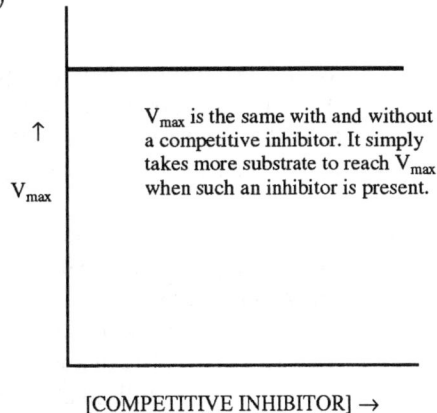

↑

V_{max}

V_{max} is the same with and without a competitive inhibitor. It simply takes more substrate to reach V_{max} when such an inhibitor is present.

[COMPETITIVE INHIBITOR] →

c)

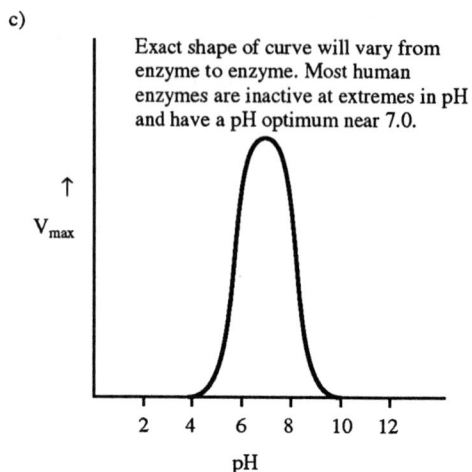

Exact shape of curve will vary from enzyme to enzyme. Most human enzymes are inactive at extremes in pH and have a pH optimum near 7.0.

$\uparrow$

V_{max}

2   4   6   8   10   12

pH

d)

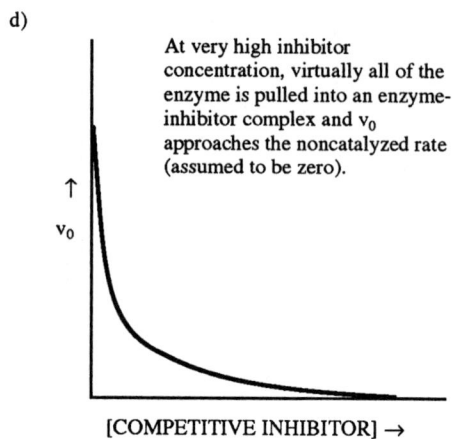

At very high inhibitor concentration, virtually all of the enzyme is pulled into an enzyme-inhibitor complex and v_0 approaches the noncatalyzed rate (assumed to be zero).

$\uparrow$

v_0

[COMPETITIVE INHIBITOR] $\rightarrow$

5.43   See Figure 5.10.

5.44

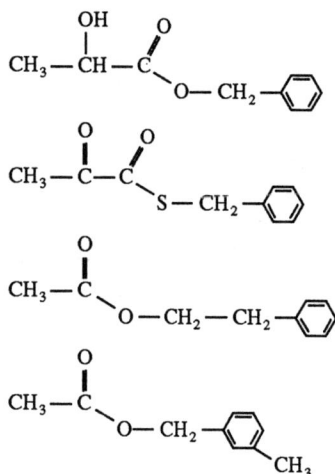

Without any details about the specific size, shape and chemical nature of the binding site, there are many possible answers. However, the best answers will be compounds with structures that are very similar to that of the substrate. A competitive inhibitor is usually one that is mistaken (by the enzyme) for a substrate.

5.45   Isozyme III. Enzymes often possess K_m's that are close to the substrate concentrations that they usually encounter, since under these conditions the rate of the catalyzed reaction is self-regulating and roughly half the enzyme molecules are working at any given time. If the K_m is far removed from the substrate concentration encountered, either the reaction exhibits poor self-regulation (often a liability) or only a small fraction of the valuable enzyme is being utilized at any instant (an inefficient arrangement).

None of the isozymes are saturated with substrate at a substrate concentration of .00009 M. Substrate concentration must be well above the K_m before an enzyme will be saturated with the substrate (see Figure 5.5). K_m, by definition, is that substrate concentration at which the enzyme is half saturated.

5.46   True. According to the concerted model, effectors function by binding to either the active or inactive form of the enzyme and pulling the bound form out of the active form-inactive form equilibrium. If inactive form is pulled out of the equilibrium, a net active to inactive conversion occurs as equilibrium is restored (Le Châtelier's principle). This explains why an effector that binds to the inactive form is a negative effector. Activators (positive effectors), in binding to the active form, pull this form out of the active-inactive equilibrium and stimulate a net inactive to active conversion.

5.47   Active sites and allosteric sites both contain stereospecific, highly discriminating binding sites that are composed of amino acid side chains arranged in a unique and precise geometry. Active sites selectively bind substrates and catalyze their conversion to products. Allosteric site bind effectors that, once bound, either stimulate or inhibit the activity of the enzyme through one or more mechanisms. In contrast to active sites, allosteric sites possess no catalytic sites or catalytic activity.

5.48   a) F could shut down its own production without inhibiting the production of K, L, or I by inhibiting Enzymes 3, 4 or 5. Unless D or E are needed for some purpose other than the synthesis of F, Enzyme 3 would be the most likely target for inhibition by Compound F. When D and E are not needed, a cell benefits by shutting down the production of D and E along with the production of F.

   b) Enzyme 6 must be the target for inhibition by K. The inhibition of any other enzyme would not lead to results consistent with the reported observations.

5.49   True. A competitive inhibitor, by definition, is one whose inhibition can be overcome by high substrate concentrations. The action of an allosteric inhibitor that reversibly pulls inactive form out of an active-inactive equilibrium (leads to the net conversion of active form to inactive form) can be overcome by high substrate concentrations, since substrate pulls the active form out of the same equilibrium (leads to the net conversion of inactive form to active form). Consider the following three reactions:

active form $\rightleftarrows$ inactive form

inhibitor + inactive form $\rightleftarrows$ inhibitor-inactive form complex

substrate + active form $\rightleftarrows$ substrate-active form complex

Think Le Châtelier's principle. As substrate is added, the bottom reaction shifts to the right, reducing the concentration of active form. As the concentration of active form drops, there is a net reversal of the top reaction which reduces the amount of inactive form. The net loss of inactive form causes the middle reaction to shift to the left freeing the enzyme from the inhibitor. At very high substrate concentrations, the substrate indirectly pulls all of the enzyme away from the inhibitor. One reaches V_{max} when all of the enzyme is bound to substrate. The same V_{max} is attained in the presence or absence of a competitive inhibitor.

5.50   A protein kinase often catalyzes the phosphorylation and activation of a second protein kinase. The second protein kinase may be less specific than the first and may catalyze the phosphorylation (and the resultant activation or inactivation) of multiple additional enzymes. Some of the additional enzymes may be other protein kinases or protein phosphatases. An initial phosphorylation can trigger a cascade of reactions that ultimately modifies the catalytic activity of many enzymes.

5.51   Hydrolases. Protein phosphatases catalyze the hydrolysis of phosphate ester links between phosphate residues and amino acid side chains in proteins.

5.52   A, b and e represent covalent modifications since they involve changes in the covalent bonding that exists within an enzyme. An H bond is not a covalent bond since it is not a product of the sharing of electrons between the bonded atoms. An H bond is an attraction between a hydrogen with a significant partial positive charge and a second atom with a appreciable partial negative charge. Denaturation, by definition, refers to the unfolding of a protein without rupturing any covalent bonds.

5.53   Since the pK_a of an alcoholic hydroxyl group is usually greater than 13, at pH 7 there will normally be over 10^6 times more of the protonated form (—OH) of this functional group than the dissociated form (—O⁻) (think Henderson-Hasselbalch equation and see Section 2.7 if a review is needed). Therefore, the serine side chain will be primarily uncharged. In contrast, the side chain of a phosphorylated serine side chain will be negatively charged at pH 7:

$$—CH_2OPO_3H_2 \rightleftarrows H^+ + —CH_2OPO_3H^- \qquad pK_a \text{ around 1}$$

$$—CH_2OPO_3H^- \rightleftarrows H^+ + —CH_2OPO_3^{2-} \qquad pK_a \text{ approximately 6}$$

When the pH is above the pK_a for a weak acid functional group, the dissociated form of the functional group is always favored.

Since phosphorylation converts a neutral side chain to a negative side chain (a rather marked chemical alteration) and changes significantly the size of a side chain, it is easy to understand how phosphorylation can either activate or inactivate an enzyme. Amino acid side chains are involved in maintaining the native conformation of an enzyme, and they are also the active components within the binding sites and catalytic sites on enzymes.

5.54   The complete inhibition of any one enzyme will totally shut down the production of compound W (assuming that the reaction catalyzed by the enzyme proceeds at an insignificant rate in the absence of a catalyst). However, the partial inhibition of one enzyme will have no effect on the rate of production of compound W if, even in the presence of inhibitor, the reaction catalyzed by the inhibited enzyme is not the rate limiting step in the conversion of Q to W. In general, the rate limiting step is the slowest step in a sequence of reactions. The rate of the rate limiting step determines the rate of the overall process; runners roped together cannot reach the finish line any more rapidly than the slowest runner in the group. All other considerations being equal, it makes sense for an allosteric effector to target the enzyme for the rate limiting step in a sequence of chemical reactions if it is the rate of the overall process that is to be regulated.

5.55   Proteinases in the stomach and intestine catalyze the hydrolysis (inactivation) of insulin when it is administered orally.

5.56
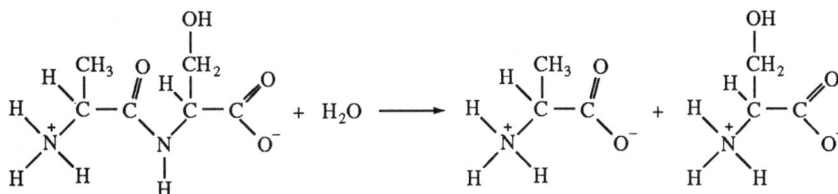

Note: at physiological pH values, carboxyl groups are dissociated and amino groups protonated.

Both digestion and proprotein activation involve the hydrolysis of peptide bonds (the reaction illustrated above).

5.57   True. Proproteins are inactive precursors of biologically active proteins. Zymogens are inactive precursors of protein enzymes.

5.58   Three. Although two cuts can yield three fragments, there must have been at least three cuts since at least one fragment was discarded during the activation of the zymogen. The three fragments within the active enzyme contain only 99 amino acid residues while the zymogen contains 107 residues. There may actually have been more than three cuts involved in the activation of the zymogen since multiple small fragments may have been created and discarded. If so, the small fragments must have collectively contained a total of 8 residues (107 minus 99).

5.59   See Table 5.1. Since each reaction in the body tends to be catalyzed by a separate enzyme, the major classes of enzymes identify the major classes of reactions.

5.60   Isomerases. D-alanine and L-alanine are isomers.

5.61   Yes. Amide and ester functional groups are similar and both functional groups are hydro-lyzed through virtually identical mechanisms. Under appropriate conditions, chymotrypsin has been found to catalyze the hydrolysis of certain esters.

5.62   At pH 1 both the histidine and the aspartic acid residues at the catalytic site would be fully protonated. Being fully protonated, they could not function as proton acceptors (general base catalysts), a role essential to the catalytic activity of chymotrypsin (see Figure 5.17). A weak acid functional group will reside almost entirely in its protonated form any time the pH is substantially below the pK_a of that functional group (Section 2.7).

5.63   The folding of a polypeptide substrate may make some tyrosine residues less accessible than others. Alternatively, neighboring side chains on the peptide substrate may make it difficult for some tyrosine side chains to fit properly into the binding pocket of chymo-trypsin or may make it difficult for a susceptible peptide bond to align itself properly at the catalytic site. The difficulty could result from any of a number of interactions between enzyme and substrate, including steric hindrance and like-charge repulsions. A proline residue in the vicinity of a tyrosine could also limit the rotational options for a peptide substrate and inhibit binding.

5.64   Normally, the amino acid whose side chain best fits into the specificity pocket "marks" the preferential site of cleavage. Since the pocket is hydrophobic and relatively deep (it nor-mally accommodates the side chains of Phe, Tyr and Trp), the most likely target would be Met in Pair #1, Val in Pair #2, and Leu in Pair #3.

5.65   Yes. Radioisotopes are commonly used as tracers in biomedical research since extremely small quantities can readily be detected. An immunoassay that employs a radioisotope is known as a radioimmunoassay (RIA). Variations on a common theme lead to multiple forms of radioimmunoassays. The RIA described in the text involves the isotopic labeling of antigen rather than antibody (immunoglobulin).

5.66   The attachment of the antibody (immunoglobulin) may change the conformation of the enzyme and destroy the precise folding required to maintain the active site. Alternatively, the presence of the antibody, a large protein itself, can block normal interactions between enzyme and substrate, even when the active site remains intact. The enzyme is most likely to remain active if the antibody is attached some distance from the active site. When cou-pling enzyme and antibody for an EIA, one must also avoid the loss of the antigen binding activity of the immunoglobulin.

5.67   Yes. Some substance from the ruptured tissue cells could rapidly destroy a specific serum enzyme. Alternatively, the damaged tissue might be the source of a serum enzyme with a short half-life. If the damaged tissue produced substantially less enzyme, serum enzyme levels could drop quickly following tissue damge. There are other possible explanations as well.

5.68   True. Since the extra Enzyme X must have come from the liver, the virus infection must have either damaged the liver or stimulated it to release Enzyme X, or both.

5.69   Asparagine is one of the 20 protein amino acids. Under normal circumstances, most of it is used as a building block to assemble peptides and proteins.

5.70   Normal cells can synthesize their own asparagine and are not dependent on an external supply. In contrast, sensitive leukemia cells must be constantly supplied with external asparagine since they have lost the ability to produce this amino acid. A cell cannot live for long without an adequate supply of each of the 20 protein amino acids. The absence of even a single amino acid makes it impossible for a cell to assemble many of the proteins that are essential for its survival. Asparaginase treatment "starves" sensitive cancer cells by destroying the required asparagine in their environment. Since asparaginase cannot enter normal cells, the asparagine they produce is not susceptible to asparaginase-catalyzed hydrolysis.

5.71   Yes. Tyrosyl-tRNA synthetase is pH sensitive since there are so many weak acid functional groups involved in catalysis (Figure 5.18). The degree of protonation of some of these functional groups changes rather markedly as pH is shifted within the physiological pH range.

5.72   The sections on the general properties of enzymes, enzyme-substrate interactions, isozymes, allosteric enzymes, protein kinases and phosphatases, zymogens, multienzyme systems, and catalytic mechanisms all deal directly with structure-function relationships. The sections on catalytic mechanisms examine this topic in greatest detail.

5.73   The structural hierarchy of a multienzyme complex is: primary structure, secondary structure, tertiary structure (can often be divided into multiple motifs and domains), and the quaternary packing of the separate polypeptides within the complex. For membrane-bound complexes, the positioning of the complex on a membrane would represent a 5th level of structural hierarchy.

5.74   Examples of the experimental basis for hypothesis (model) construction are described in the two sections on catalytic mechanisms (Sections 5.16 and 5.17). Also, see the last paragraph in Sections 5.9 and 5.10.

6

Carbohydrates

	(A)		(B)		(C)		(D)
	OH		H		O‖		O‖
1	H—C—H	1	H—C—OH	1	C—H	1	C—H
2	HO—C—H	2	C=O	2	H—C—OH	2	HO—C—H
3	C=O	3	H—C—OH	3	H—C—OH	3	HO—C—H
4	H—C—OH	4	HO—C—H	4	HO—C—H	4	HO—C—H
5	H—C—OH		H	5	H—C—OH	5	H—C—H
6	HO—C—H			6	H—C—OH		OH
7	HO—C—H				H		
	H						
	L		D		D		L

↑	↑	↑	↑
enantiomers	enantiomers	enantiomers	enantiomers
↓	↓	↓	↓

(A)	(B)	(C)	(D)
CH₂OH	CH₂OH	O‖C—H	O‖C—H
H—C—OH	C=O	HO—C—H	H—C—OH
C=O	HO—C—H	HO—C—H	H—C—OH
HO—C—H	CH₂OH	H—C—OH	H—C—OH
HO—C—H		HO—C—H	CH₂OH
H—C—OH		CH₂OH	
CH₂OH			

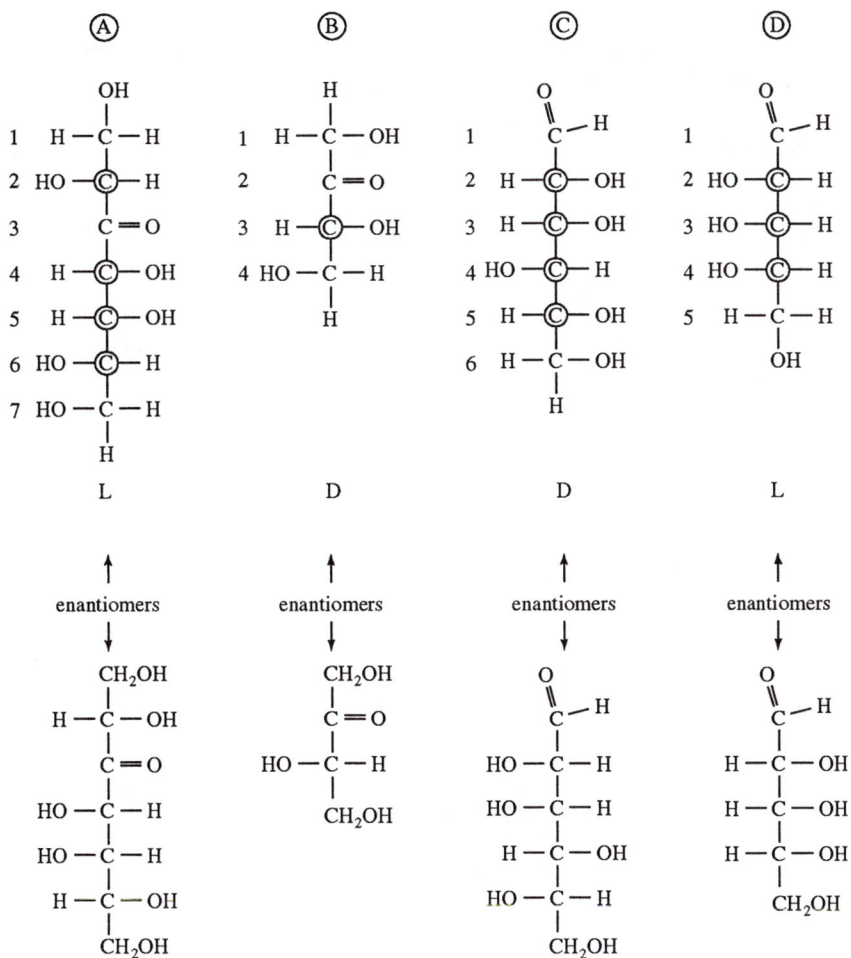

Note: If a carbon is not a chiral center, it does not matter how one orients attached groups around it.

e) C (2nd from right) and D (far right) are aldoses; they are aldehydes.

A (far left) is a heptose; it contains 7 carbons.

A and B are ketoses; they are ketones.

None are oligosaccharides. They are all monosaccharides. Oligosaccharides are formed by joining 2 to 11 monosaccharides.

D is a pentose; it contains 5 carbons.

6.2   Anomers are not enantiomers; they are not nonsuperimposible mirror images.

Anomers are diastereomers; they are stereoisomers that are not enantiomers.

Anomers are stereoisomers; they differ in the spatial arrangement of atoms but have the same molecular formulas and the same connectedness of atoms.

Anomers are not structural (constitutional) isomers; their atoms have the same connectedness.

6.3

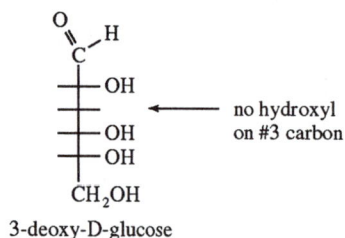

3-deoxy-D-glucose

← no hydroxyl on #3 carbon

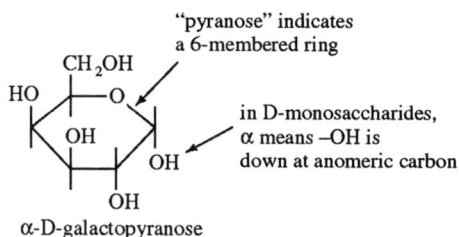

"pyranose" indicates a 6-membered ring

in D-monosaccharides, α means –OH is down at anomeric carbon

α-D-galactopyranose

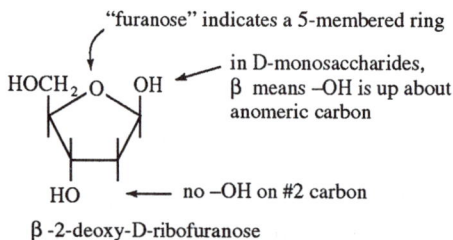

"furanose" indicates a 5-membered ring

in D-monosaccharides, β means –OH is up about anomeric carbon

← no –OH on #2 carbon

β-2-deoxy-D-ribofuranose

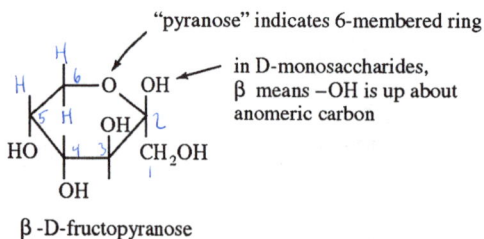

"pyranose" indicates 6-membered ring

in D-monosaccharides, β means –OH is up about anomeric carbon

β-D-fructopyranose

6.4   Addition of the #4 hydroxyl across the carbonyl group leads to two furanose forms (α and β). Addition of the #6 hydroxyl generates two 7-membered ring forms (α and β). The addition of the #2 and #3 hydroxyls across the carbonyl group generates 3 and 4-membered rings, respectively, that are disfavored as a consequence of ring strain. The cyclic forms exist in equilibrium with the open chain form.

6.5

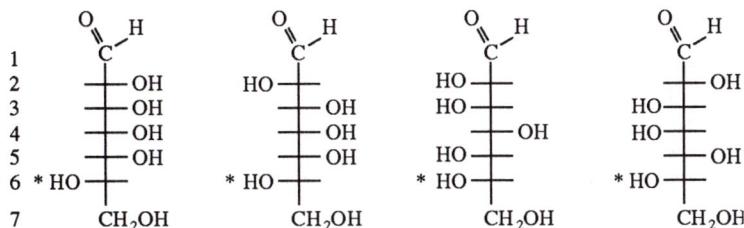

An aldoheptose contains 7 carbons with the #1 carbon part of an aldehyde functional group. To be an L-monosaccharide, the highest numbered chiral center (identified with a *) must have an L configuration (–OH to the left).

There are 12 other possible answers. For a monosaccharide, the total number of stereoisomers (including the monosaccharide) is 2^n where n is the number of chiral centers. For aldoheptoses, there are 5 chiral centers and 32 possible stereoisomers. Half of these are L-monosaccharides, the other half D-monosaccharides.

6.6

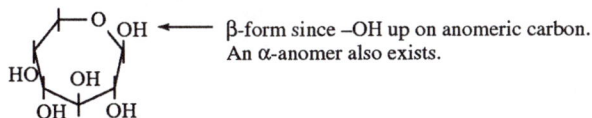

β-form since –OH up on anomeric carbon.
An α-anomer also exists.

6.7

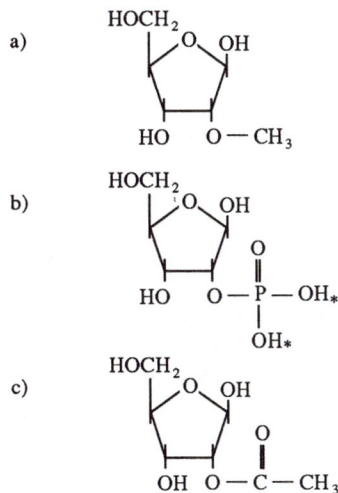

a)

b)

c)

* Exists predominately in a dissociated form a physiological pH.

6.8   A reducing sugar can exist in a readily oxidizable open chain form that possesses a free carbonyl group. Nonreducing sugars cannot exist in a carbonyl-containing form.

6.9   Glucose.

6.10

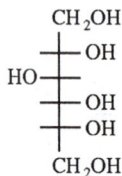

Aldehydes are reduced to primary alcohols

6.11

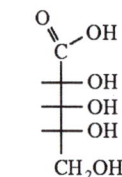

Aldehydes are oxidized to carboxylic acids

6.12

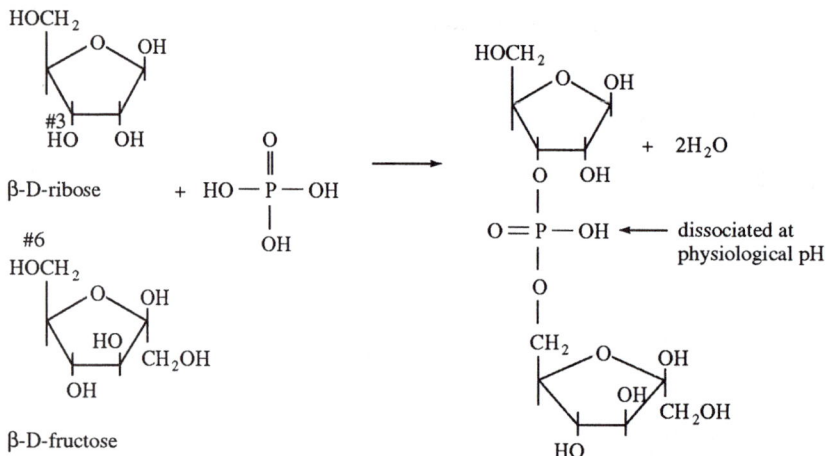

β-D-ribose   + HO—P—OH

β-D-fructose

6.13   The reaction of glucose with hemoglobin to form Amadori products is a bimolecular reaction whose rate increases as the concentration of reactants increases. For this reason, a rise in red blood cell glucose concentrations triggers the more rapid production and more rapid accumulation of glycosylated hemoglobin. As blood glucose levels are elevated, glucose levels also increase within red blood cells (the only cells that produce hemoglobin). If this were not the case, glycosylated hemoglobin could not be used to monitor hyperglycemia.

6.14 No. The free α-amino groups in hemoglobin react with CO_2 to form carbamates, a transport form of CO_2 (Section 4.9). Once these free amino groups have reacted with glucose they are no longer available to bind and carry CO_2. Glycosylation also changes the conformation and chemical character of hemoglobin to such an extent that it is no longer able to interact normally with its effectors (2,3-BPG, H^+, etc.) and O_2.

6.15

HOCH$_2$ ... O ... OH #2 ... OH ... α ... #1 ... HOCH$_2$ ... O ... OH OH ... O

This residue can exist in an α, β (as shown) or open chain form. The open chain option makes this a reducing sugar.

HOCH$_2$ ... O ... #1 α ... OH OH ... O ... #5 CH$_2$ ... O ... OH ... OH OH

This residue can exist in an α (as shown), β or open chain form. The open chain option makes this a reducing sugar.

HOCH$_2$ ... O ... #1 α ... OH OH ... HOCH$_2$ ... O ... O ... #1 β ... HO OH

A nonreducing sugar since no carbonyl-containing open chain form is possible

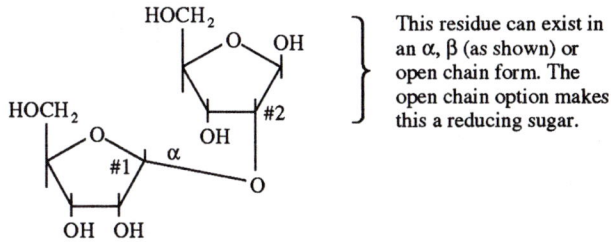

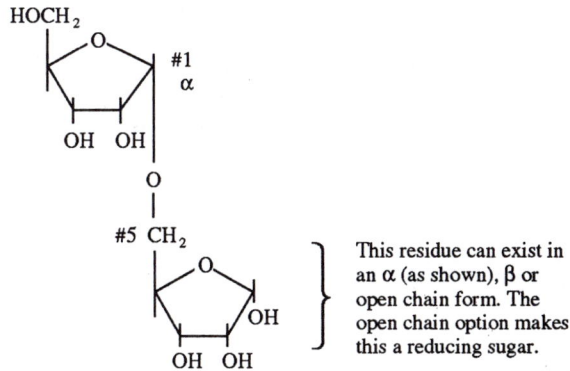

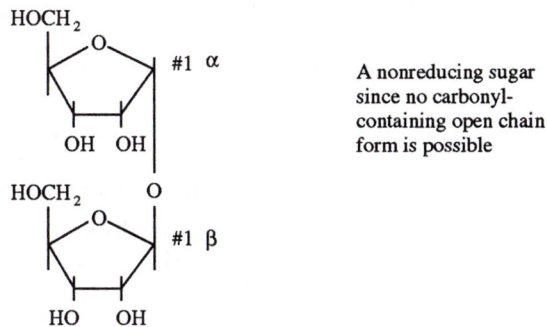

The term "sugar" is used loosely in this problem. The disaccharides shown may or may not be sweet tasting.

6.16 Starch: many glucose; starch is a polymer of glucose. In all of the oligo and polysaccharides examined in this chapter (including starch), the monosaccharides are joined through hydrolyzable glycosidic (acetal) links.

Fructose: no hydrolysis; monosaccharides, by definition, cannot be hydrolyzed.

Sucrose: glucose and fructose; sucrose is a disaccharide containing these two monosaccharides.

Glucose: no hydrolysis; a monosaccharide.

Glycogen: many glucose; glycogen is a polymer of glucose.

Chitin: many glucosamines plus an equal number of acetic acid molecules; chitin is a polymer of N-acetylglucosamine. Both acetal links (hold monomers together) and amide bonds (hold the acetyl groups to nitrogens) are hydrolyzable.

Cellulose: many glucose; cellulose is a polymer of glucose.

Lactose: galactose and glucose; lactose is a disaccharide containing these two monosaccharides.

6.17 Starch: many glucose.

Glycogen: many glucose.

Lactose: glucose and galactose.

In each case, digestion consists of hydrolysis (see Exercise 6.16).

6.18 Both linear and branched homopolymars can be produced from each of the listed monosaccharides. A linear homopolymer is a polymer in which many copies of the same monomer ("homo" means same) are joined in a long continuous (linear) chain. Cellulose and amylose are linear homopolymers. A branched homopolymer is one in which monomer chains branch off of other monomer chains. Glycogen and amylopectin are examples. Any monosaccharide capable of forming glycosidic (acetal) links can be used to generate both linear and branched homopolymers. However, living organisms are not genetically programmed to synthesize all of the conceivable polymers.

6.19 The enzymes that an organism is genetically programmed to produce determine what reactions occur within it. The human body, for example, has genes that code for the construction of the enzymes involved in the assembly of glycogen, but it lacks genes for some of the enzymes involved in the production of cellulose. Virtually every reaction that occurs within a living organism is catalyzed by a unique and highly specific enzyme.

6.20   Immunoglobulins (immune response; Table 4.1, Section 4.11), fibrinogen (plays a central role in blood clotting), some hormones (chemical messengers; Section 10.3), some enzymes (catalysts; Chapter 5), some transport proteins, some receptors (for hormones, neurotransmitters, etc.), and some structural proteins are all glycoproteins. The antigens responsible for blood groups are also glycoproteins. Additional examples will be encountered throughout the text.

6.21   Hyaluronic acid, chondroitin sulfates, and heparin all exist as polyanions at physiological pH. These compounds have carboxyl groups or sulfate groups, or both, scattered along their polymer chains.

6.22   Primarily H bonds and salt bridges. Chondroitin sulfates are too polar to participate in hydrophobic interactions. Although present, van der Waals interactions are relatively weak.

6.23   Hyaluronic acid is a lubricant around bone joints and a structural element in connective tissue.

6.24   The human body is not genetically programmed to produce the catalysts necessary to hydrolyze the glycosidic links that exist between the monomers within cellulose and chitin. The body does carry genes for the synthesis of amylases, highly specific enzymes involved in the digestion (hydrolysis) of starch and glycogen.

6.25   Active transport, facilitated diffusion, and free diffusion. Only active transport requires energy. This energy is usually provided through the coupled cleavage of ATP (Chapters 7 and 8).

6.26   A symbiotic relationship exists between cows and their intestinal flora (those bacteria found in their intestine). These bacteria produce the enzymes, known as cellulases, that allow cows to digest (hydrolyze) cellulose. When the bacteria are destroyed with an antibiotic, a cow rapidly runs out of cellulases. Cows are not genetically programmed to produce their own cellulose-degrading enzymes.

6.27   No. Diffusion is the movement of a substance from a region of higher to a region of lower concentration. If Compound X is present at higher concentrations in blood than in the intestine, this compound will tend to diffuse from the blood into the intestine rather than vice versa. Within the human body, diffusion is commonly blocked by membranes on other barriers.

6.28   Fiber increases the rate at which fecal material moves through the intestine. This reduces the time that bacterially-produced carcinogens linger in the large intestine. The fiber may also adsorb some of the carcinogens and reduce the rate at which they are absorbed into the bloodstream.

6.29   False. Cellulose is a nonfattening carbohydrate since the body cannot digest this carbohydrate to get at the calories that it contains. The body must be able to digest and absorb a compound before it can use that substance to build fats.

6.30   C

6.31   Intravenous glucose is more likely to cause hyperglycemia. When compared to oral glucose, which stimulates GIP production in the stomach, intravenous glucose leads to a reduction in the amount of insulin released into the blood by the pancreas. As blood insulin levels drop, blood sugar levels rise because insulin stimulates the removal of glucose from the blood.

6.32   The sugar, itself, is no better. In fruit, however, the sugar is packaged with a multitude of nutrients, including a variety of vitamins and minerals. Refined sucrose provides only carbon and calories, no essential nutrients.

6.33   Glucose must bind to a taste receptor to trigger a sweet taste. Starch is unable to bind to such receptors.

6.34   a) Glycosidic (acetal) bonds.

   b) Primarily H bonds. Cellulose has many hydroxyl (-OH) groups that participate in interchain H bonding.

   c) Glycosidic bonds.

   d) Primarily H bonds between the hydroxyl groups in fructose and the hydrogens and oxygen in water.

7

Nucleotides and
Some Related Compounds

7.1

a)

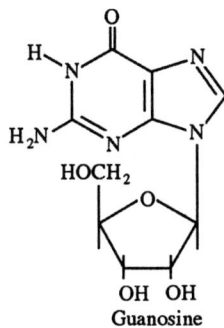

Guanosine

b)

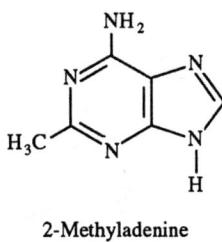

2-Methyladenine

c)

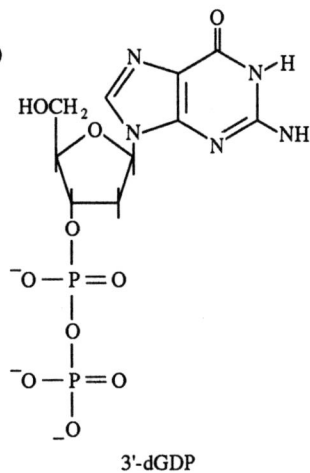

3'-dGDP

d)

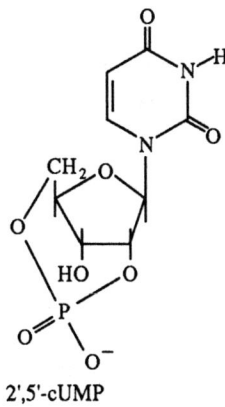

2',5'-cUMP

e)

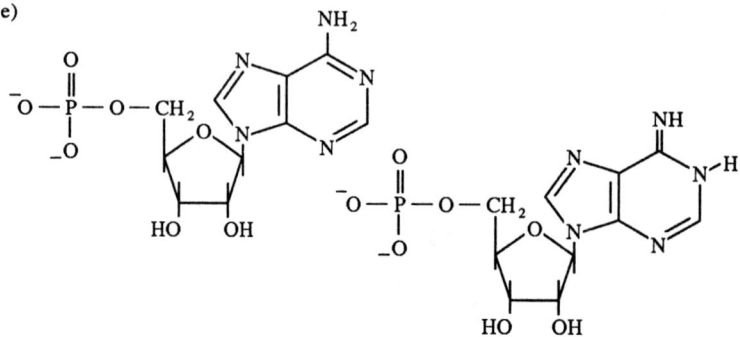

Two Tautomers of AMP

f)

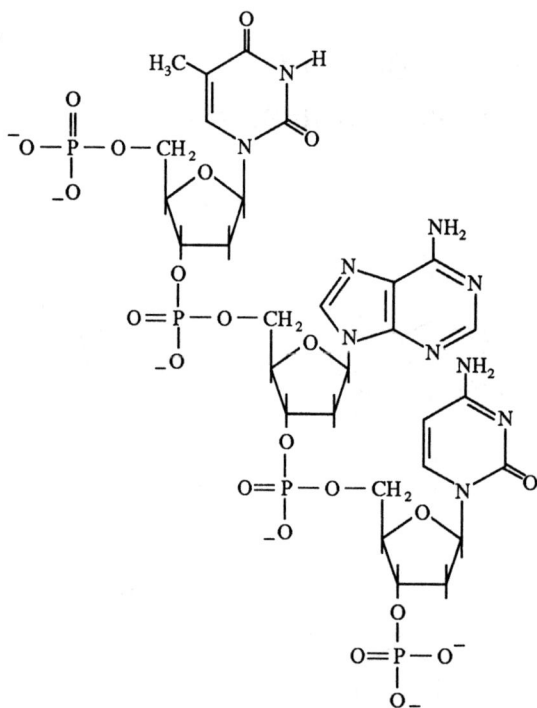

dpTpApCp

g)

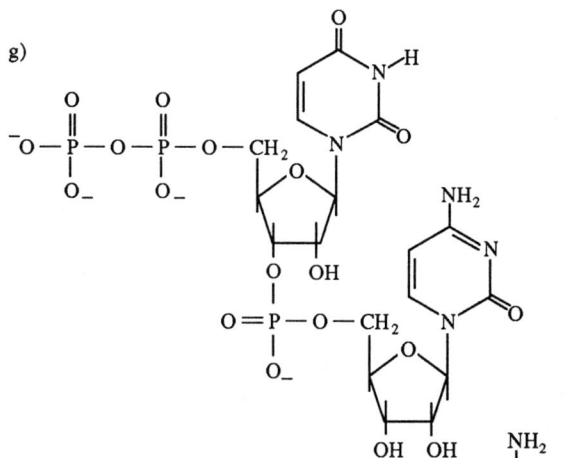

h)

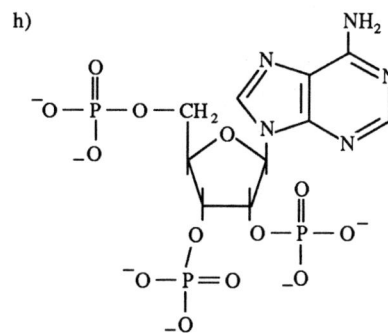

Adenosine -2',3',5'-trisphosphate

ppUpC

i)

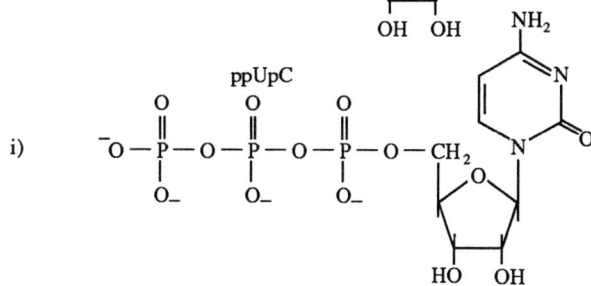

5' CTP

j)

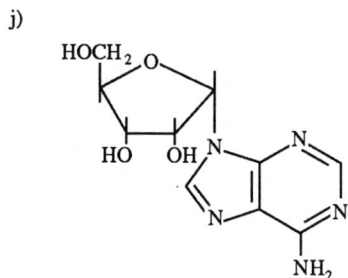

Ribose joined to adenine
through an α-N-glycosidic link

This answer assumes that the ribose is
attached to the #9 nitrogen in the purine
ring. Other options do exist when one
starts working with nucleosides that are
not of biological origin.

7.2   E = ester (phosphate)
        A = anhydride (phosphate)
        G = glycosidic
        All of the hydrolyzable bonds are ester, anhydride or glycosidic bonds.

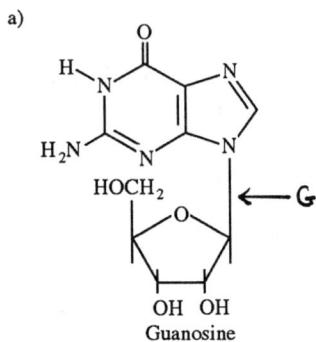

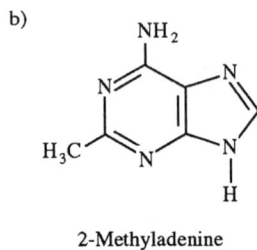

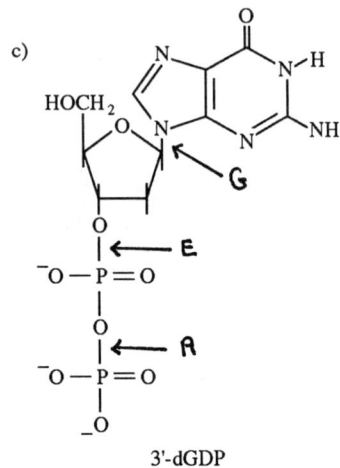

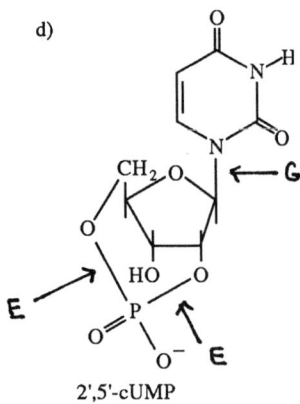

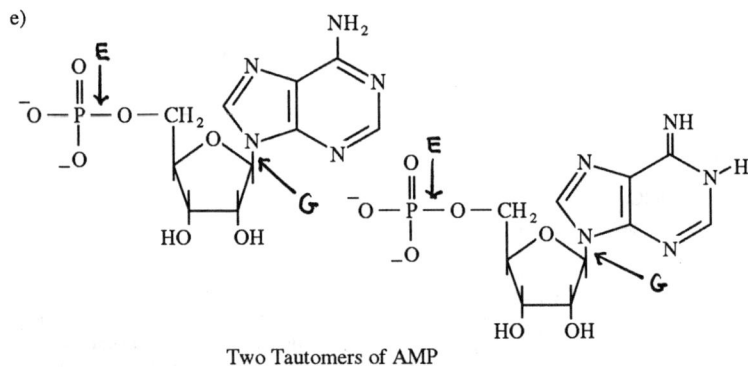

a)

HOCH₂

OH OH
Guanosine

b)

2-Methyladenine

c)

HOCH₂

3'-dGDP

d)

2',5'-cUMP

e)

Two Tautomers of AMP

f)

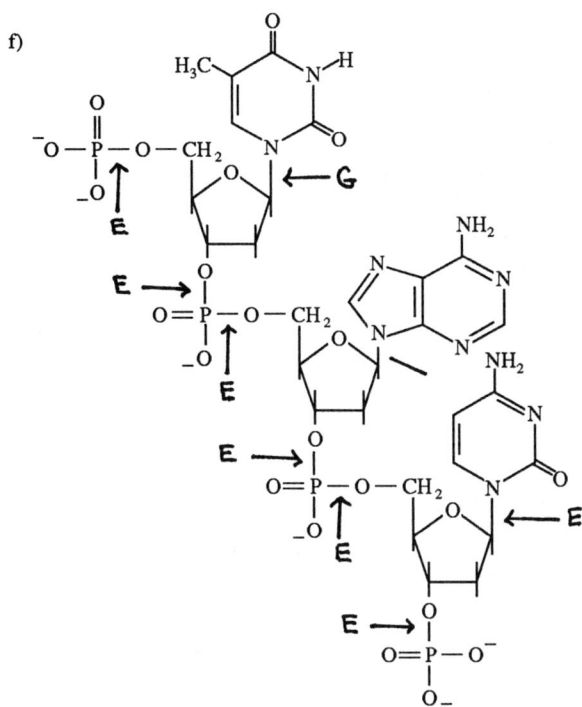

dpTpApCp

g)

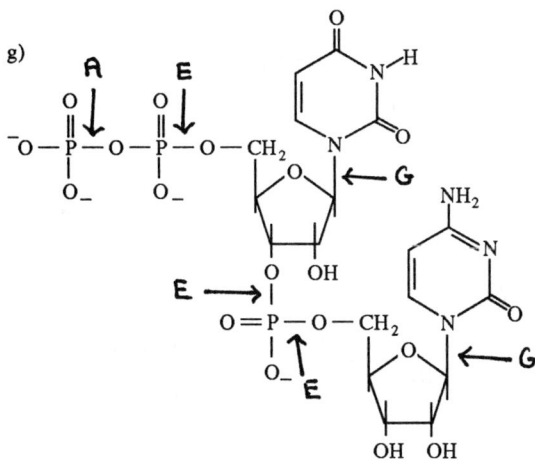

ppUpC

h)

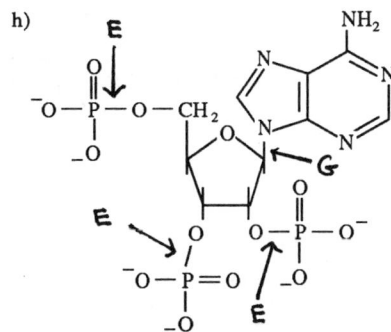

Adenosine -2',3',5'-trisphosphate

i)

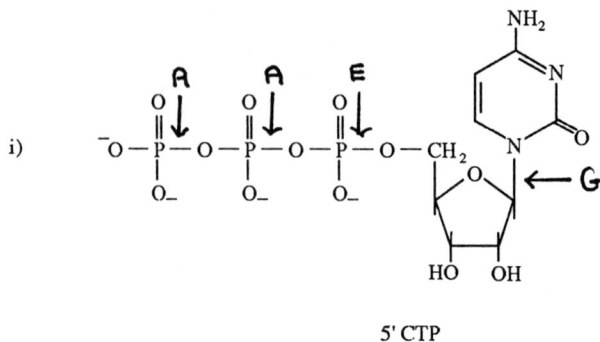

5' CTP

j)

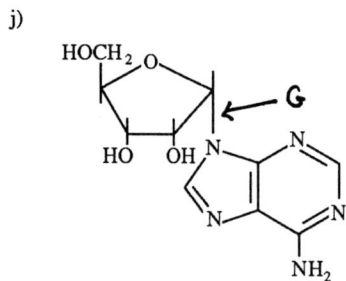

Ribose joined to adenine
through an α-N-glycosidic link

This answer assumes that the ribose is
attached to the #9 nitrogen in the purine
ring. Other options do exist when one
starts working with nucleosides that are
not of biological origin.

7.3

a)

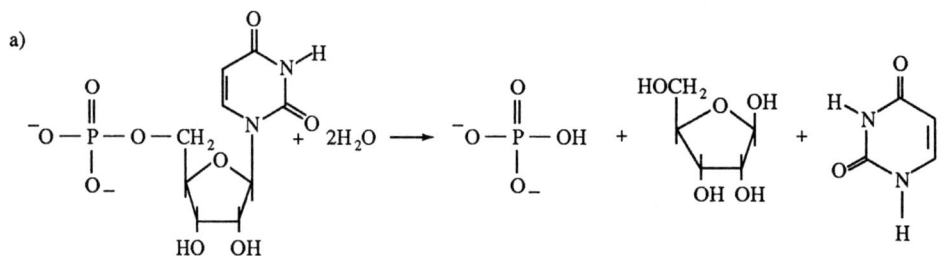

b)

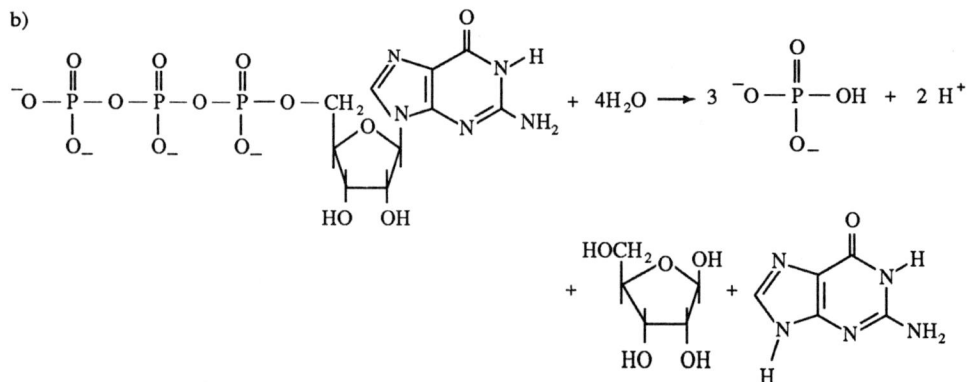

c)

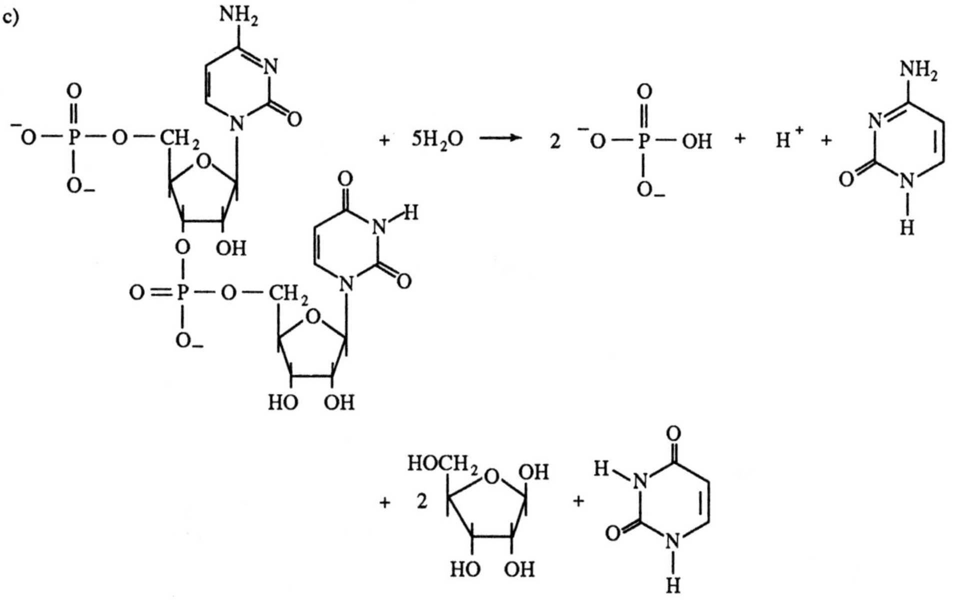

NH_2

$+ 5H_2O \longrightarrow$ 2 $^-O-P-OH$ $+$ H^+ $+$

NH_2

$+$ 2 ... $+$

d)

$+ 3H_2O \longrightarrow$ 2 $^-O-P-OH$ $+$ H^+ $+$

$+$

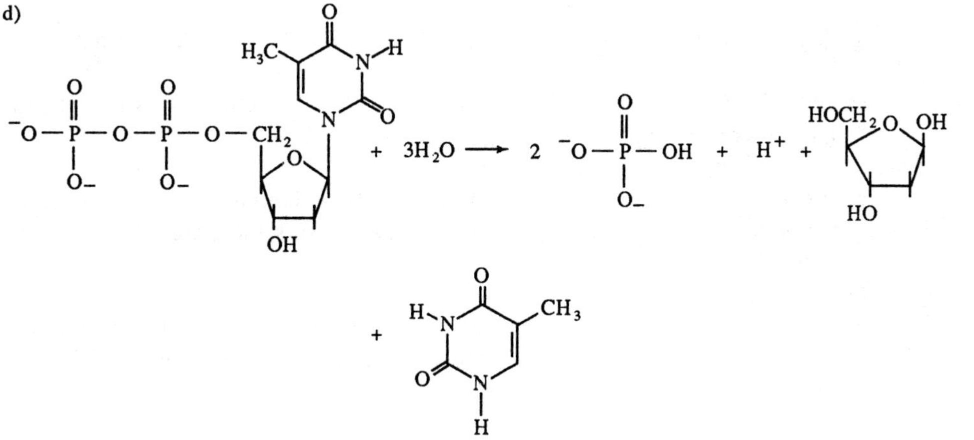

e)

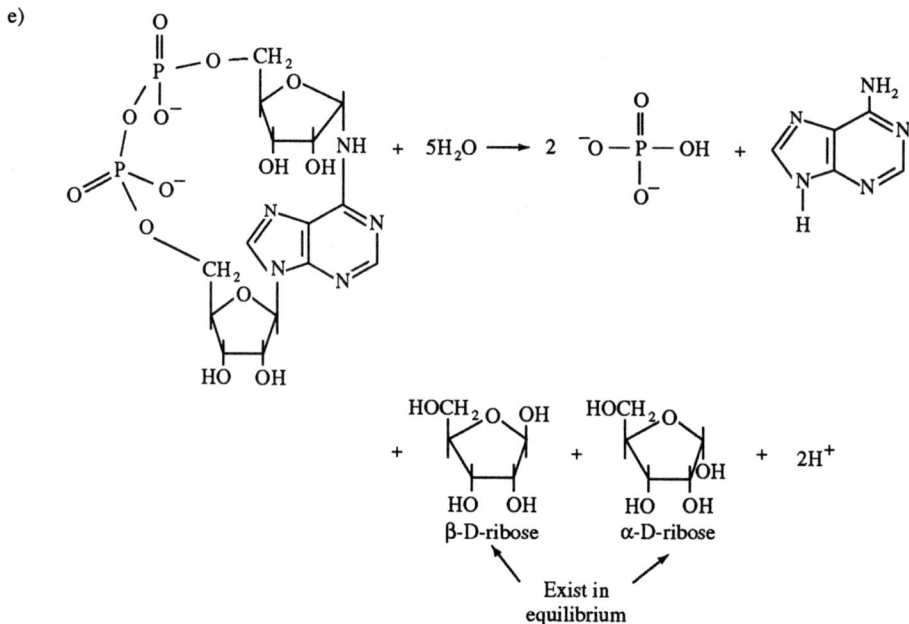

β-D-ribose

α-D-ribose

Exist in
equilibrium

7.4   a) The net charge on AMP is –2.0 at pH 7.0. See Problem 7.7; AMP is a nucleotide.

b) The net charge on pCpUpGp is –6 at pH 7.0. See Problem 7.10; exactly the same calculations are applicable.

c) The net charge on adenosine is zero at pH 7.0. Although the purine ring contains weak acid functional groups and can exist in charged forms, these are disfavored, minor forms at pH 7.0 (Section 7.1).

d) This octapeptide will have a net charge of –9 at pH 7.0. Each internal phosphate residue will bare a –1 charge while the 5' phosphate residue has a –2 charge (all calculations to the nearest half of a charge unit).

7.5   Protein kinases catalyze the phosphorylation of proteins. Serine, threonine and tyrosine side chains are common targets for the phosphorylations (Section 5.13).

7.6

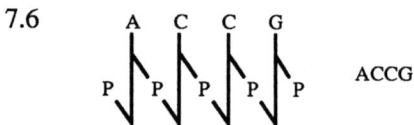

ACCG

7.7

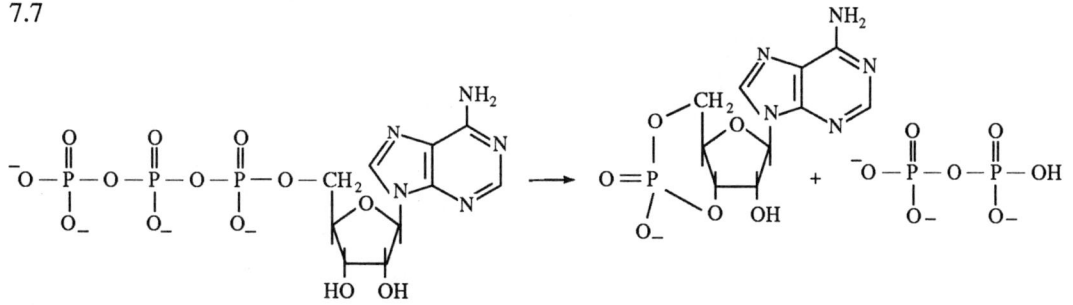

8

Bioenergetics

8.1   The energy in our food can, though the food chain, be traced to photosynthetic organisms that transform sunlight into chemical energy in carbohydrates. The oxidation of dietary organic compounds releases this chemical energy. This is the major source of the energy used by the human body.

Most of the energy used by our society comes from the burning of fossil fuels (coal, oil, and natural gas). The chemical energy within fossil fuel comes from the organic compounds in ancient photosynthetic organisms that served as starting materials for the production of this fuel.

8.2   Chemical energy in a battery is transformed into electrical energy used to create a spark. The spark initiates the combustion of gasoline, a process that releases additional chemical energy. Part of this chemical energy is converted to kinetic energy within pistons, gears, drive shafts, and the automobile as a whole. Most of this chemical energy is transformed into heat, another form of kinetic energy.

8.3   Most CO_2 is eliminated from the body through the lungs as one exhales. A significant fraction of this CO_2 is transported to the lungs covalently attached to hemoglobin (Table 4.6). CO_2 reacts with N-terminal α-amino groups in hemoglobin to form carbamates.

8.4   Water is a by-product of the oxidation of organic fuel molecules:

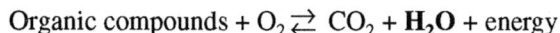

Organic compounds + $O_2 \rightleftarrows CO_2 + \textbf{H}_2\textbf{O}$ + energy

Some desert animals produce all of the water they need through this process.

8.5   True. The chemical and physical changes that occur within cells do not create or destroy energy; they only lead to energy transformations.

8.6   False. The conditions specified correspond to internationally agreed upon standard conditions. Few reactions are at equilibrium (have a $\Delta G^{\circ\prime} = 0$) under these conditions. None of the reactions encountered in this chapter, for example, have $\Delta G^{\circ\prime} = 0$.

8.7   No math is required to answer the first two questions, since the ratio of Y to X at equilibrium is, by definition, equal to the equilibrium constant:

$$K_{eq} = \frac{[Y]}{[X]} = \frac{3}{1}$$

The equilibrium ratio will remain the same if the initial concentration of X is doubled, because an equilibrium constant is independent of the initial concentration of reactants and products.

To solve the last part of this question:

$$K_{eq} = 3.0 = \frac{[Y]}{[X]} = \frac{[Y]}{0.064}$$

$[Y] = (3.0)(0.064) = 0.19$ M

8.8 $\Delta G^{\circ} = -RT \ln K_{eq} = -8.315$ J/mol $\bullet$ K $\times$ 298 K $\times \ln 3.0 = -2.7 \times 10^3$ J/mol

8.9   False. $\Delta G_p = \Delta G^{\circ\prime} + RT \ln \dfrac{[C]^c[D]^d}{[A]^a[B]^b}$

RT ln $\dfrac{[C]^c[D]^d}{[A]^a[B]^b}$ can, in theory, have any value. ΔG_p will be negative whenever

RT ln $\dfrac{[C]^c[D]^d}{[A]^a[B]^b}$ has a absolute negative value greater than the positive value of $\Delta G^{\circ\prime}$.

8.10   True. When a reaction is at equilibrium, the driving force is zero; there will be no spontaneous net reaction and no work can be accomplished by the reaction. If the product to reactant ratio is far above that at equilibrium, there is a large driving force for the reverse reaction. When product to reactant ratio is much less than that at equilibrium, there is a large driving force for the forward reaction.

8.11   False. $\Delta G_p = \Delta G^{\circ\prime} + RT \ln \dfrac{[C]^c[D]^d}{[A]^a[B]^b}$

A large negative ΔG_p indicates that a reaction is far from equilibrium under physiological conditions and that a net forward reaction is required to reach equilibrium. Alone, it provides no information about the value of the equilibrium constant. The concentrations of A, B, C, and D in the ΔG_p expression are physiological concentrations, not equilibrium concentrations.

8.12 $\Delta G_p = \Delta G^{\circ\prime} + RT \ln \dfrac{[C]^c[D]^d}{[A]^a[B]^b}$

$\Delta G_p = -20,900 \text{ J/mol} + 8.315 \text{ J/mol} \cdot K \times 298 \text{ K} \ln \dfrac{(1.0 \times 10^{-2}\text{M})(2.0 \times 10^{-3}\text{M})}{1.0 \times 10^{-6}\text{M}}$

Notice that, by convention, $[H_2O]$ is not included in $RT \ln \dfrac{[C]^c[D]^d}{[A]^a[B]^b}$.

$\Delta G_p = -20,900 + 2,480 \ln 20 = -20,900 + 7,400 = -13,500 \text{ J/mol}$

A net forward reaction is required to reach equilibrium since ΔG_p is negative.

8.13   a) 2, 4 and 6 are spontaneous, since $\Delta G^{\circ\prime}$ is negative.

   b) 1 and 5 have a K_{eq} less than 1, since the $\Delta G^{\circ\prime}$ is positive.

   c) 4 and 6 represent the hydrolysis of a high energy compound, since the $\Delta G^{\circ\prime}$ is more negative than -25 kJ/mol.

   d) One cannot determine relative rates from $\Delta G^{\circ\prime}$ values.

8.14   a) $\Delta G^{\circ\prime}$ is approximately -62 kJ/mol, since two phosphate anhydride bonds are hydrolyzed. The $\Delta G^{\circ\prime}$ is around -31 kJ/mol for the hydrolysis of each anhydride bond.

   b) $\Delta G^{\circ\prime}$ is approximately -31 kJ/mol, since a single phosphate anhydride bond is hydrolyzed.

8.15   No. K_{eq} is a ratio of concentrations, and concentrations cannot have negative values. If K_{eq} is less than one, the $\ln K_{eq}$ is negative, but K_{eq} is still positive.

8.16   True. Suppose that compound A has a free energy of 600 kJ/mol and compound B a free energy of 800 kJ/mol. The $\Delta G^{\circ\prime}$ for the hydrolysis of compound A is -42 kJ/mol, but compound B cannot be hydrolyzed. Compound A is, by definition, a high energy compound (its $\Delta G^{\circ\prime}$ for hydrolysis is <-25 kJ/mol) while compound B is not. However, compound B possesses more total free energy than compound A. Without a precise definition, the term "high energy compound" can be misinterpreted.

8.17   The $\Delta G^{\circ\prime}$ for the former reaction is -32 kJ/mol while the $\Delta G^{\circ\prime}$ for the latter reaction is -31 kJ/mol. Thus, the hydrolysis to yield $AMP + PP_i$ gives a slightly greater driving force. For many purposes, this difference is insignificant. When ATP is hydrolyzed to $AMP + PP_i$ within a cell, the PP_i is usually hydrolyzed to help "pull" the reaction (LeChâtelier's principle).

8.18   Since the hydrolysis of each ATP makes the coupled process more negative by about 31 kJ/mol, a minimum of 5 ATP would need to be hydrolyzed; 150 divided by 31 = 4.84 or 5 (to the next larger whole number). The coupled hydrolysis of 5 ATP would make the $\Delta G^{\circ\prime}$ for the process -5 kJ/mol [$(-31)(5) + 150$].

8.19   Protein X + ATP $\rightleftarrows$ ADP + Protein X – P

8.20   More. It is identical to ATP except it has one additional phosphate residue and one additional phosphate anhydride bond. The hydrolysis of an anhydride bond releases considerable energy.

8.21   Less. 3′,5′-ABP has two phosphate ester bonds but no phosphate anhydride bonds. In contrast, ADP has one ester bond and one anhydride bond. The hydrolysis of an anhydride bond releases more energy than the hydrolysis of an ester bond.

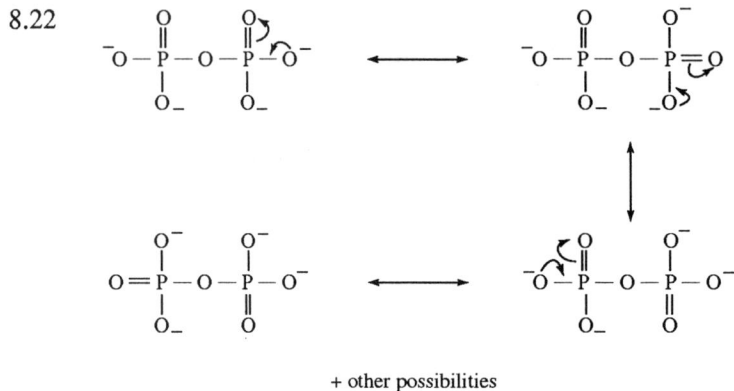

8.22

+ other possibilities

8.23 $\Delta G^{\circ\prime} = -RT \ln K_{eq}{}'$

$$\ln K_{eq} = \frac{\Delta G^{\circ\prime}}{-RT} = \frac{4400 \text{ J/mol}}{-8.315 \text{ J/mol} \cdot K \times 298} = -1.8$$

$K_{eq} = $ antiln $-1.8 = 0.17$

If $\Delta G_p = +25$ kJ/mol, a net reverse reaction is required to reach equilibrium. A negative ΔG_p signals that a net forward reaction is required to reach equilibrium. When $\Delta G_p = 0$, a reaction is at equilibrium under physiological conditions.

8.24 $\Delta G_p = \Delta G^{\circ\prime} + RT \ln \dfrac{[C]^c [D]^d}{[A]^a [B]^b}$

Therefore, $\Delta G_p = \Delta G^{\circ\prime}$ when $RT \ln \dfrac{[C]^c [D]^d}{[A]^a [B]^b} = 0$

This occurs whenever $\dfrac{[C]^c [D]^d}{[A]^a [B]^b} = 1$

8.25   Table 8.7 provides no information about rates or ΔG_p's. $\Delta G^{\circ\prime}$ is totally independent of rate while ΔG_p is determined by reactant and product concentrations as well as $\Delta G^{\circ\prime}$:

$$\Delta G_p = \Delta G^{\circ\prime} + RT \ln \frac{[C]^c [D]^d}{[A]^a [B]^b}$$

8.26 ΔG_p will become more negative: the hydrolysis of PP_i will reduce the concentration of products relative to the concentration of reactants and pull the reaction further from equilibrium, since a net forward reaction is initially required to reach equilibrium. In general, ΔG_p becomes more negative (it may still be positive if it was initially positive) as the ratio of products to reactants decreases.

8.27   False. $\Delta G = \Delta H - T\Delta S$. Although ΔH is negative for exothermic reactions, ΔG is positive whenever $T\Delta S$ has an absolute negative value greater than that of ΔH.

8.28   True. When entropy increases, ΔS is positive. When ΔS is positive, $T\Delta S$ makes a negative contribution to ΔG, since $\Delta G = \Delta H - T\Delta S$. As ΔG becomes more negative, the driving force increases.

8.29   Increases. In the gaseous state, water molecules have more freedom of movement (exhibit greater randomness of movement) than they do when in the liquid state. Entropy is a measure of the randomness or disorder in a system.

8.30   Reduced. Incorporation into glucose involves the addition of hydrogens and the loss of an oxygen, two events that lead to a gain of electrons by the carbon in CO_2. Reduction is the gain of electrons.

8.31   Reactions a, c and f are redox reactions. Reactions b and d are group transfer reactions that do not involve any transfer of electrons between atoms. Reaction e is an isomerization reaction.

a) $4\,Fe$ + $3\,O_2$ $\rightleftharpoons$ $2\,Fe_2O_3$

    reducing agent   oxidizing agent
       oxidized   reduced
   electron donor   electron acceptor

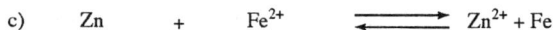

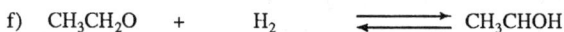

c) Zn + Fe^{2+} $\rightleftharpoons$ $Zn^{2+} + Fe$

    reducing agent   oxidizing agent
       oxidized   reduced
   electron donor   electron acceptor

f) CH_3CH_2O + H_2 $\rightleftharpoons$ CH_3CHOH

   oxidizing agent   reducing agent
       reduced   oxidized
  electron acceptor   electron donor

8.32   Transformations a, b, d and f are oxidations. Transformations c and e are reductions. The answer to f can be obtained from Table 8.9. You must know the structure of the given compounds or classes of compounds to answer the remaining questions. Think "organic chemistry" or consult an organic chemistry text. Each transformation involves the addition of H or removal of O or both (leads to reduction) or involves the removal of H or addition of O or both (leads to oxidation).

8.33   E is the reduction potential for a test half reaction at any concentrations of acceptor and donor when $[H^+] = 1$ M and $E°$ is the point of reference:

$$E = E° + \frac{RT}{n\mathfrak{F}} \ln \frac{[\text{electron acceptor}]}{[\text{electron donor}]}$$

$E°$ is the reduction potential under standard conditions (298 K, 101.3 kPa partial pressure for gaseous reactants and products, and 1 M concentration of all nongaseous reactants and products).

E′ is the reduction potential for a test half reaction at any concentrations of acceptor and donor when $[H^+] = 10^{-7}$ M and $E°′$ is the point of reference:

$$E′ = E°′ + \frac{RT}{n\mathfrak{F}} \ln \frac{[\text{electron acceptor}]}{[\text{electron donor}]}$$

$E°′$ is the reduction potential under standard biological conditions (298 K, 101.3 kPa partial pressure for all gaseous reactants and products, pH 7.0, 55.5 M H_2O, and 1 M concentration of other nongaseous reactants and products).

8.34   True.

$$E = E° + \frac{RT}{n\mathfrak{F}} \ln \frac{[\text{electron acceptor}]}{[\text{electron donor}]}$$

$\ln \frac{[\text{electron acceptor}]}{[\text{electron donor}]}$ becomes more positive (it may still be negative if negative originally) as the ratio of electron acceptor to electron donor increases.

8.35   None of these are spontaneous under standard conditions. In each case, the electron acceptor in the reactants (NADP$^+$ in a, NAD$^+$ in b, glutathione in c, and pyruvate in d) is the oxidizing agent. The electron donor in the reactants is the reducing agent. A reaction is spontaneous if the electron acceptor in the reactants has a greater electron affinity (a larger $E°′$) than the acceptor in the products. The relative electron affinities can be determined from Table 8.9. In reaction a, for example, NADP$^+$ (the acceptor in the reactants) has the same electron affinity ($E°′ = -0.320$) as NAD$^+$ (the acceptor in the products, $E°′ = -0.320$); the reaction is not spontaneous. Similar comparisons for the other reactions lead to the same conclusion.

8.36 $E' = E^{o'} + \dfrac{RT}{n\mathscr{F}} \ln \dfrac{[\text{electron acceptor}]}{[\text{electron donor}]}$

$E' = -0.185 + \dfrac{0.026}{2} \ln \dfrac{10 \times 10^{-3}}{2 \times 10^{-3}} = -0.185 + 0.021 = 0.164 \text{ V}$

8.37   The reducing agents are on the right in each half reaction equation while the oxidizing agents are to the left. The lower the $E^{o'}$ for a half reaction, the more powerful its reducing agent (because the oxidizing agent has a lower affinity for electrons and its reduced form will give up electrons more readily). All of the reducing agents below $FADH_2$ (in flavoproteins) in Table 8.9 are more powerful than $FADH_2$ (in flavoproteins). This includes malate, lactate, ethanol, $FADH_2$ (free), reduced glutathione, and so on.

The higher the $E^{o'}$ for a half reaction, the more powerful its oxidizing agent. Fumarate, ubiquinone, the cytochromes (Fe^{3+}), Fe^{3+}, and O_2 are all more powerful oxidizing agents than FAD (in flavoproteins), because they all have a greater electron affinity then FAD (in flavoproteins).

8.38 $\Delta E^{o'} = 0.771 - 0.031 = 0.740 \text{ V}$

$\Delta G^{o'} = -n\mathscr{F}\, \Delta E^{o'}$

$\Delta G^{o'} = -(2)\,(96.48 \text{ kJ/V} \cdot \text{mol})(0.740 \text{ V}) = \mathbf{-143\ kJ/mol}$

$\Delta G^{o'} = -RT \ln K_{eq}'$

$\ln K_{eq}' = \dfrac{\Delta G^{o'}}{-RT}$

$\ln K_{eq}' = \dfrac{-143000 \text{ J/mol}}{-8.315 \text{ J/mol} \cdot \text{K} \times 298 \text{ K}} = 57.7$

$K_{eq}' = \text{antiln } 57.7 = \mathbf{1.16 \times 10^{25}}$

8.39 $\Delta G^{o'} = -n\mathscr{F}\, \Delta E^{o'}$

$\Delta E^{o'} = \dfrac{\Delta G^{o'}}{-n\mathscr{F}} = \dfrac{23.5 \text{ kJ/mol}}{-(1)(96.48 \text{ kJ/V} \cdot \text{mol})}$

$\Delta E^{o'} = -0.244 \text{ V}$

Lipids and Biological Membranes

9.1

$$CH_2 - O - \overset{O}{\overset{\|}{C}} - (CH_2)_{16} - CH_3$$
$$CH - O - \overset{O}{\overset{\|}{C}} - (CH_2)_{16} - CH_3$$
$$CH_2 - O - \overset{O}{\overset{\|}{C}} - (CH_2)_{16} - CH_3$$

$$CH_2 - O - \overset{O}{\overset{\|}{C}} - (CH_2)_{16} - CH_3$$
$$CH - O - \overset{O}{\overset{\|}{C}} - (CH_2)_7 - CH = CH - (CH_2)_5 - CH_3$$
$$CH_2 - O - \overset{O}{\overset{\|}{C}} - (CH_2)_{16} - CH_3$$

$$CH_2 - O - \overset{O}{\overset{\|}{C}} - (CH_2)_{16} - CH_3$$
$$*CH - O - \overset{O}{\overset{\|}{C}} - (CH_2)_{16} - CH_3$$
$$CH_2 - O - \overset{O}{\overset{\|}{C}} - (CH_2)_7 - CH = CH - (CH_2)_5 - CH_3$$

$$CH_2 - O - \overset{O}{\overset{\|}{C}} - (CH_2)_{16} - CH_3$$
$$*CH - O - \overset{O}{\overset{\|}{C}} - (CH_2)_7 - CH = CH - (CH_2)_5 - CH_3$$
$$CH_2 - O - \overset{O}{\overset{\|}{C}} - (CH_2)_7 - CH = CH - (CH_2)_5 - CH_3$$

$$CH_2 - O - \overset{O}{\overset{\|}{C}} - (CH_2)_7 - CH = CH - (CH_2)_5 - CH_3$$
$$CH - O - \overset{O}{\overset{\|}{C}} - (CH_2)_{16} - CH_3$$
$$CH_2 - O - \overset{O}{\overset{\|}{C}} - (CH_2)_7 - CH = CH - (CH_2)_5 - CH_3$$

$$CH_2 - O - \overset{O}{\overset{\|}{C}} - (CH_2)_7 - CH = CH - (CH_2)_5 - CH_3$$
$$CH - O - \overset{O}{\overset{\|}{C}} - (CH_2)_7 - CH = CH - (CH_2)_5 - CH_3$$
$$CH_2 - O - \overset{O}{\overset{\|}{C}} - (CH_2)_7 - CH = CH - (CH_2)_5 - CH_3$$

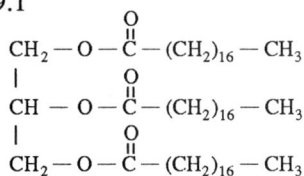

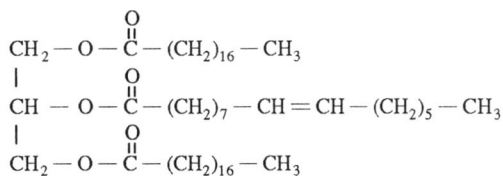

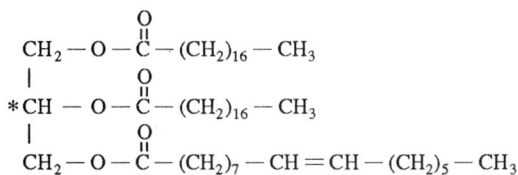

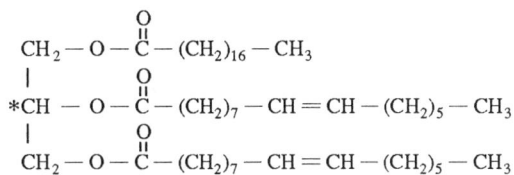

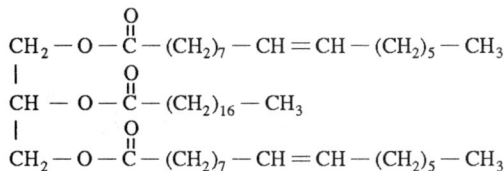

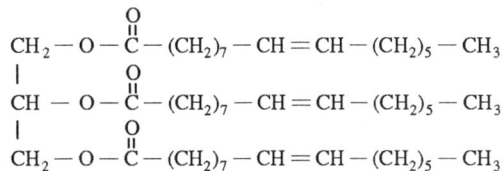

The starred (*) carbons are chiral centers that can have either a D or an L configuration. Only one of the two possible stereoisomes is assembled by living organisms as a consequence of the stereospecificity of the enzymes involved.

9.2   Vegetable oils tend to be unsaturated while animal fats are generally saturated. The carbon-carbon double bonds in vegetable oils make them more susceptible to oxidation, the major process that causes fats to go rancid. Even without knowledge of this latter fact, one would predict that unsaturated fats, containing more functional groups, would be chemically more reactive that saturated ones.

9.3   A triacylglycerol containing unsaturated fatty acid residues tends to be a liquid at room temperature, while triacylglycerols with saturated fatty acid residues are usually solids at the same temperature.

There are many possible answers including:

$$CH_2 - O - \overset{\overset{\displaystyle O}{\|}}{C} - (CH_2)_6 - (CH_2 - CH = CH)_2 - (CH_2)_4 - CH_2$$

$$CH - O - \overset{\overset{\displaystyle O}{\|}}{C} - (CH_2)_2 - (CH_2 - CH = CH)_5 - CH_2 - CH_3 \qquad \text{probably a liquid fat}$$

$$CH_2 - O - \overset{\overset{\displaystyle O}{\|}}{C} - (CH_2)_7 - CH = CH - (CH_2)_5 - CH_3$$

$$CH_2 - O - \overset{\overset{\displaystyle O}{\|}}{C} - (CH_2)_{14} - CH_3$$

$$CH - O - \overset{\overset{\displaystyle O}{\|}}{C} - (CH_2)_{16} - CH_3 \qquad \text{probably a solid fat}$$

$$CH_2 - O - \overset{\overset{\displaystyle O}{\|}}{C} - (CH_2)_{12} - CH_3$$

9.4   The multiple possible answers include:

$$\overset{1}{CH_3} - \overset{2}{CH_2} - \overset{3}{CH} = \overset{4}{CH} - \overset{5}{CH_2} - CH_2 - CH_2 - CH_2 - CH_2 - CH_2 - CH_2 - C\overset{\displaystyle O}{\underset{\displaystyle OH}{}}$$

$$\overset{1}{CH_3} - \overset{2}{CH_2} - \overset{3}{CH} = \overset{4}{CH} - \overset{5}{CH_2} - CH = CH - CH_2 - CH_2 - CH_2 - CH_2 - C\overset{\displaystyle O}{\underset{\displaystyle OH}{}}$$

$$\overset{1}{CH_3} - \overset{2}{CH_2} - \overset{3}{CH} = \overset{4}{CH} - \overset{5}{CH_2} - CH = CH - CH_2 - CH = CH - CH_2 - C\overset{\displaystyle O}{\underset{\displaystyle OH}{}}$$

To be an ω-3 carboxylic acid, an acid must have a double bond between the #3 and #4 carbons when one numbers from the noncarboxyl end. Only a small fraction of the imaginable ω-3 carboxylic acids are fatty acids (building blocks for fats).

9.5   Vegetable oils are esters and alkenes while car oils are, for the most part, complex mixtures of saturated hydrocarbons. At the high temperatures within an automobile engine, esters and alkenes are much more reactive than saturated hydrocarbons. Vegetable oils would be rapidly converted to substances unable to provide the necessary lubrication for the pistons.

9.6 CO_2 and H_2O plus energy. These are the final products generated during the complete oxidation of any organic compound that contains only carbon, oxygen and hydrogen. It does not matter whether the oxidation is carried out within a living organism or within a test tube.

9.7   Like associates with like, a major theme of biochemistry. As nonpolar substances move through the body, they will tend to associate with any lipids deposits they encounter. Adipose tissue and liver are the major fat depots in the body.

9.8   a) 10 g carbohydrate × 4 kcal/g = 40 kcal
             2 g protein × 4 kcal/g = 8 kcal
             2 g fat × 9 kcal/g = 18 kcal
             Total kcal = 66 kcal

18/66 × 100 = 27% of calories from fat

b) 5 g carbohydrate × 4 kcal/g = 20 kcal
         5 g protein × 4 kcal/g = 20 kcal
         5 g fat × 9 kcal/g = 45 kcal
         Total kcal = 85 kcal

45/85 × 100 = 53 % of calories from fat

9.9   Normal roles of fat (predominantly triacylglycerols but also contains smaller amounts of mono- and diacylglycerols):
     store fuel (energy)
     pad vital organs
     insulate
     store fat-soluble vitamins
     signal transduction by some diacylglycerols

Potential risks of excess consumption of saturated fats:
     cancer
     heart disease
     obesity and its associated health risks

9.10   Yes. Fats are the major source of the essential fatty acids, those fatty acids that the body requires but cannot produce. The essential fatty acids are precursors of eicosanoids and a variety of other compounds.

9.11   Serine, threonine and tyrosine each contain a side chain hydroxyl group that can participate in ester formation. Cysteine, with a thiol group in its side chain, can form thioesters.

9.12   Table 9.4

Some Glycerophospholipids

Subclass	Structure	Alcohol Esterified to Phosphatidate
Phosphatidates	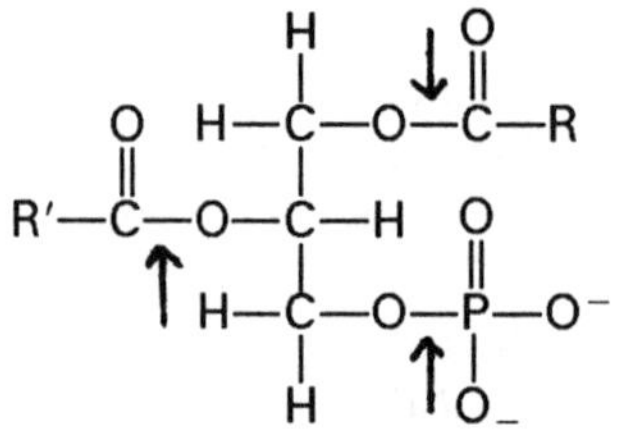	Arrows identify the hydrolyzable bonds
Phosphatidylcholines (lecithins)	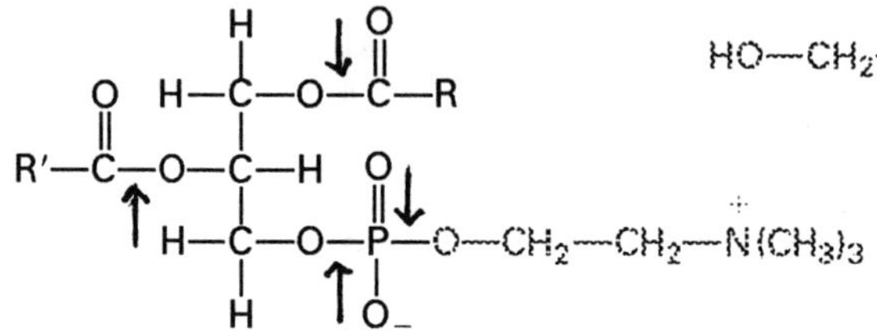	$HO-CH_2-CH_2-\overset{+}{N}(CH_3)_3$ (choline)
Phosphatidylethanolamines (cephalins)	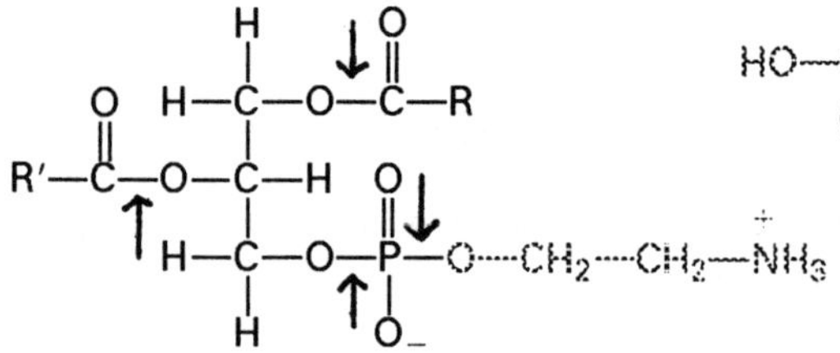	$HO-CH_2-CH_2-\overset{+}{N}H_3$ (ethanolamine)
Phosphatidylinositols	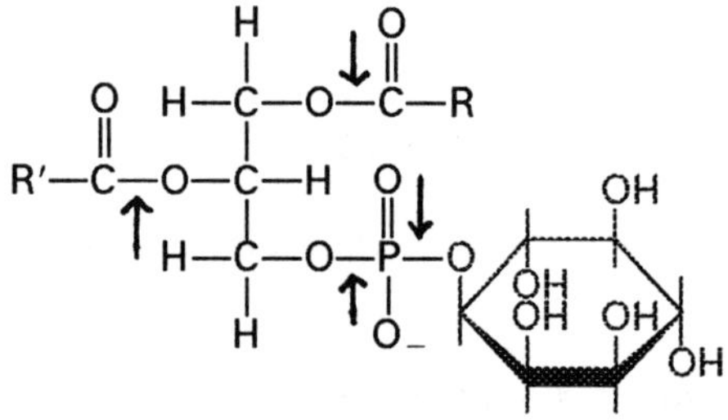	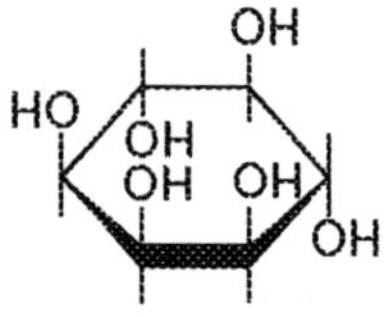 (myo-inositol)
Phosphatidylserines	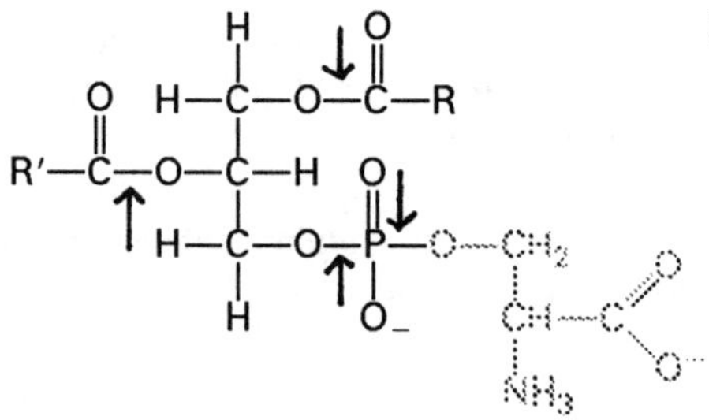	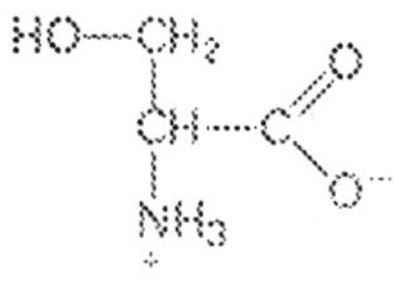 (serine)

9.13   Although most fat molecules (predominantly triacylglycerols) contain three ester links, these links are not polar enough to constitute a polar head. An amphipathic compound must have a *major* polar component (the polar head) plus a major nonpolar component. The polar head in most amphipathic lipids is charged.

9.14   A sphingomyelin is: a lipid; a phospholipid; an amphipathic compound; a common membrane component; and chiral.

9.15   A cerebroside is: a lipid; a glycolipid; an amphipathic compound; a common membrane component; and chiral.

A cholesterol is: a lipid; a steroid; a common membrane component; and chiral. It does have a polar (due to the -OH group) and nonpolar side, but it is not highly amphipathic.

A monoacylglycerol is: a lipid;, a minor component of fat; and chiral.

PGE_2 is: a lipid; an eicosanoid; and chiral.

9.16   Increase. Carbohydrates are highly polar, primarily as a consequence of their multiple hydroxyl groups.

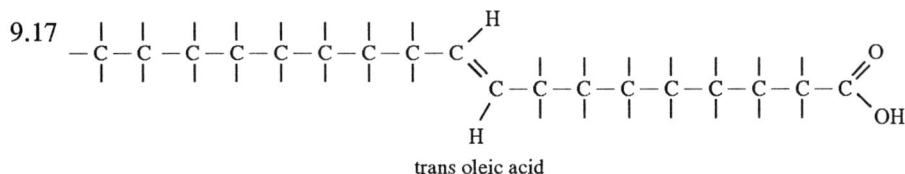

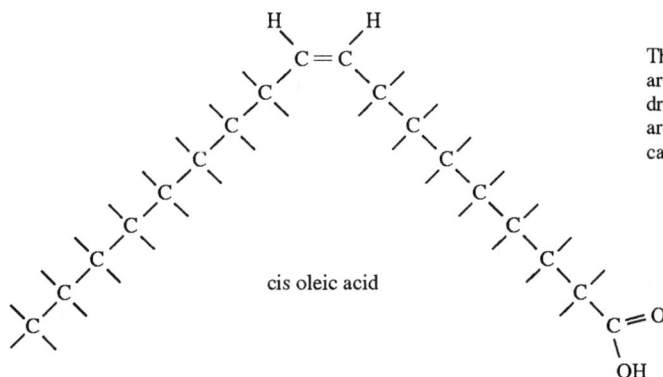

9.17

trans oleic acid

cis oleic acid

The true differences in shape are exaggerated in these drawings. Many conformations are possible about each carbon-carbon single bond.

Cis oleic acid is a kinked molecule that does not pack well with like neighbors. Consequently, separate molecules of the *cis* isomer cling together less tightly than separate molecules of the *trans* isomer, and they can more readily be pulled apart. Melting involves a partial pulling apart of neighboring molecules. Thus, the *cis* isomer melts more easily, at the lower temperature, while the *trans* isomer requires more heat input to separate the molecules, and melts at the higher temperature.

9.18   Unique glucose-containing cerebrosides differ in their fatty acid residues. The many possible answers include:

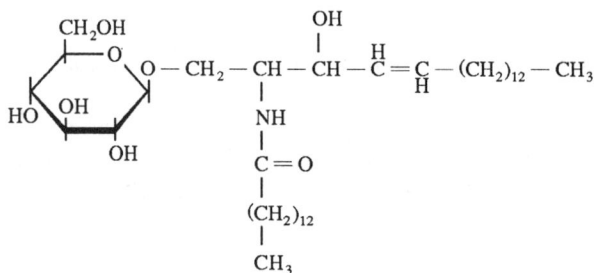

and

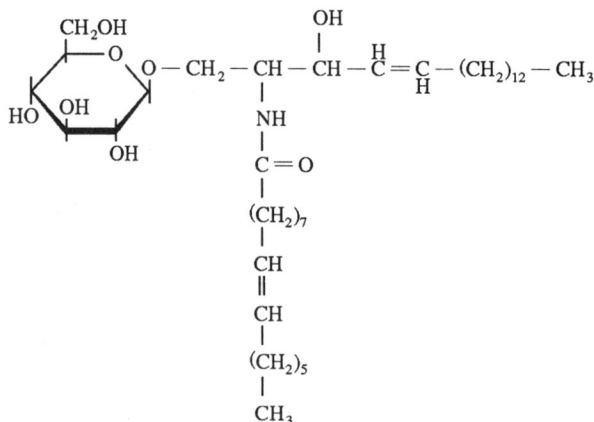

Note: It is possible for glucose to be joined to the sphingosine residue through an α-glycosidic link rather than a β-glycosidic link (as shown).

9.19   Glucose, stearic acid and sphingosine. Glucose is joined to sphingosine through a hydrolyzable acetal (glycosidic) link, while steric acid is bonded to sphingosine through a hydrolyzable amide link.

9.20   Taurocholate is saponifiable. Taurocholate is the only steroid in Figure 9.16 that contains a readily hydrolyzable link, an amide link. Alcohols, phenols, ketones and alkenes cannot be hydrolyzed.

9.21

Function of Steroids	‡Steroid that Serves Listed Function
Membrane Components	Cholesterol
Vitamin Precursors	Cholesterol
Hormones	Cortisol, testosterone, progesterone, estradiol
Aid in Digestion	Taurocholate
Antibiotics	Squalamine

‡Examples discussed in Chapter 9. Not a comprehensive list.

9.22   Geometric isomerism requires restricted rotation about one or more carbon-carbon bonds. Due to its ring system, cholesterol does have restricted rotation about some if its carbon-carbon bonds and, although not shown in Figure 9.16, geometric isomerism is possible. However, cholesterol has a specific configuration as a consequence of the stereospecificity of the biosynthetic enzymes responsible for its production.

9.23   If the pK_a is 4, the salt form will predominate at pH 7. Whenever the pH is above the pK_a for a weak acid, one has more conjugate base than weak acid (Section 2.7). The ionic bond in the salt makes it more soluble in water, a highly polar solvent (like dissolves like).

9.24   Arachidonic acid

9.25   Like associates with like. One side of most apolipoproteins extends into the aqueous environment surrounding lipoprotein particles, while the other side is buried in the nonpolar core of these particles. The side extending into the aqueous environment must be coated with polar and charged side chains in order for lipoproteins to be soluble in blood. The amino acid residues on the surface of that side of the protein buried in the nonpolar core must contain nonpolar side chains. Alanine, cysteine, glycine, isoleucine, leucine, methionine, phenylalanine, proline, tryptophan, and valine possess side chains that are nonpolar or borderline polar/nonpolar (Table 3.6).

9.26   Yes. Different classes contain unique apoproteins or unique combinations of apoproteins (Table 9.5).

9.27   Like associates with like. Free cholesterol contains a highly polar hydroxyl group which makes one side of this steroid significantly more polar than the other side and gives it some amphipathic properties. When a fatty acid is esterified to cholesterol, the hydroxyl group on cholesterol is eliminated and a long hydrocarbon chain is added. Both of these changes make the esters significantly less polar and less amphipatic than cholesterol.

9.28   Waxes have no significant polarity and, consequently, no tendency to mix with polar water molecules. Restated, nonpolar waxes tend to exclude polar water molecules.

9.29   Peripheral proteins possess a polar side that has difficulty passing through the nonpolar core of a lipid bilayer. The same central theme appears time and time again; polar and nonpolar tend to exclude one another.

9.30   Detergents are amphipathic compounds that interact with and disperse both polar and nonpolar substances. Detergents do not usually act by breaking covalent bonds, they simply compete with membrane components for noncovalent interactions. An exception would be certain enzyme-containing detergents, but these are not used to disrupt membranes.

9.31   Cholesterol molecules are primarily bound to their receptors by hydrophobic forces and H bonds. The hydroxyl group in cholesterol participates in H bonding while its larger hydrocarbon component participates in hydrophobic interactions.

9.32   Polar and nonpolar tend to exclude one another. Polar will not dissolve in nonpolar. Carbohydrates are highly polar while the inside of a lipid bilayer is nonpolar.

9.33   A doubling of transporters doubles V_{max} but has no effect on K_m. If one doubles the number of transporters, twice as many substrates can be transported per minute, but the concentration of substrate that half saturates the transporters will not change. Similarly, in an enzyme catalyzed reaction, V_{max} is proportional to enzyme concentration while K_m is independent of this variable (Sections 5.6 and 5.7).

9.34   Small K_m. When substrate levels are usually low, transporters can only work efficiently if half saturated at low substrate concentrations. If the K_m is high, very few of the transporters will be working at any instance. K_m is the substrate concentration that half saturates the transporter; it is usually an inverse measure of how tightly a substrate binds its transporter. A review of Section 5.6 would be beneficial at this point.

9.35   Both facilitated diffusion and active transport involve membrane-spanning protein transporters that interact with specific substrates and usually exhibit saturation kinetics. In contrast to facilitated diffusion, active transport requires energy and can move a substrate against a concentration gradient.

   All active transport involves a substrate-specific, membrane-spanning protein transporter that requires energy and exhibits saturation kinetics. Primary active transport uses a primary source of energy, usually ATP, a photon of light, or a redox reaction. Secondary active transport employs the energy associated with a concentration gradient that was created with primary active transport.

9.36   No. The amount of energy consumed is determined by the mechanism of the transport process. The consumption of a large amount of energy does not necessarily indicate that the rate of energy consumption and the rate of transport is rapid. The rate of transport is primarily determined by substrate concentration, the K_m of the transporter, inhibitor concentration (if any), temperature, the activity of participating enzymes (if any), and the activation energy for the conformational change that usually moves the substrate from one side of the membrane to the second. Thus, the factors that determine the amount of energy consumed differ from those that determine rate of transport.

Transporter A is involved in active transport, since it requires energy. Transporter B must be involved in facilitated diffusion. Under appropriate conditions, facilitated diffusion can proceed more rapidly than active transport and vice versa.

9.37   Since it takes approximately 5.5 turns of the helix to traverse the bilayer, this lipid complex must be roughly .56 nm/turn $\times$ 5.5 turns = 3.1 nm thick.

9.38   Since the $GluT_1$ tunnel must be polar in order to accommodate the polar glucose molecule, one would expect the tunnel to be lined with amino acid residues bearing polar or charged side chains. Serine, threonine, aspartic acid, glutamic acid, asparagine, glutamine, tyrosine, lysine, histidine, and arginine are the amino acids with polar or charged side chains a physiological pH values.

9.39   Since glucose can flow in both directions across $GluT_2$, the rate of net transport during facilitated diffusion is determined by the magnitude of the concentration gradient. As blood glucose levels increase following a typical meal, the concentration gradient across the plasma membrane of the liver cell decreases. The flow of glucose out of liver cells remains constant (assuming internal concentration constant) while the rate of glucose transport into these cells increases. This leads to a reduction in the rate of net transport of glucose out of the liver.

9.40   Greater. Since mannose has a lower affinity for $GluT_1$, more mannose will be required to half saturate the transporter. By definition, K_m is the substrate concentration that half saturate a transporter.

9.41   True. Facilitated diffusion, like free diffusion, is, by definition, the flow of a substance from higher to lower concentration. In contrast to diffusion, flow against a concentration gradient requires energy.

9.42   Step 2, that step where ATP is hydrolyzed and the transporter is phosphorylated. Cleavage of the anhydride link in ATP releases considerable energy.

9.43   The rate of Na^+ transport is independent of Na^+ concentration at concentrations where Na^+ is saturating. It is also independent of Na^+ concentrations at K^+ concentrations where the second half of the transport cycle (the transport of K^+) is rate limiting. When K^+ transport is the slowest part of the cycle, a change in the rate of the first half of the cycle will have no impact on the rate of movement through the entire cycle.

9.44   Entropy-linked potential energy it the free energy that a system possesses as a consequence of its entropy. Part or all of this energy is released when a system moves to a more random state. A concentration gradient contains entropy-linked potential energy because it possesses less entropy (less disorder) than a system where concentration is uniform (more disordered).

9.45   Both channels and transporters are membrane-spanning proteins involved in the translocation of select substances across membranes. While channel transport is always by diffusion, transporter-linked movement may be against a concentration gradient. Channels, in contrast to transporters, are often gated and normally exhibit nonsaturation kinetics. Transporters, in contrast to channels, contain specific binding sites for those substances they transport and normally exhibit saturation kinetics.

9.46   A channel exhibits saturation kinetics under conditions where a substrate moves through the channel at a slower rate than it enters the channel through diffusion. Under these conditions, further substrate would lead to no increase in transport rate and the channel would be operating at V_{max}.

9.47   Endocytosis—a clathrin-coated pit in the plasma membrane invaginates and then pinches off to form an internalized coated vesicle that eventually opens to release its contents into some part of the cell.

Exocytosis—an internal, membrane-enclosed vesicle fuses with the plasma membrane of a cell to release its contents outside the cell.

Potocytosis—a tiny caveolin-coated cave (a caveola) in the plasma membrane closes up to form a tiny plasma membrane-attached vesicle. Substances within sealed caveolae move into the cytosol through caveolae membranes.

Transcytosis—the movement of sealed caveolae from site to site on the plasma membrane.

10

Vitamins, Hormones and
an Introduction to Metabolism

10.1   Fat-soluble vitamins are stored in fat-containing cells (primarly hepatocytes and adipocytes), but the body has no mechanism for storing the water-soluble vitamins.

10.2   The amide, ester and N-acetal functional groups can be hydrolyzed.

10.3   Boiling leads to chemical alterations in most vitamins. Many of the alterations are due to hydrolysis and oxidation.

10.4   Niacin, pantothenic acid, biotin and folic acid are carboxylic acids. Virtually all carboxylic acids have pK_a values well below pH 7. Whenever the pH is above the pK_a for a carboxylic acid, the acid will exist predominantly in its negatively-charged, dissociated form (Section 2.7).

10.5   Thiamine, riboflavin, pyridoxine, pantothenic acid, ascorbic acid, cobalamin, vitamin A and vitamin D are alcohols. Primary alcohols can be oxidized to aldehydes and carboxylic acids. Secondary alcohols can be oxidized to ketones. All alcohols can be dehydrated and used as building blocks for esters, including carboxylic acid esters and phosphoric acid esters. Alcohols can also be converted to ethers, acetals and hemiacetals. Although alcohols undergo additional reactions as well, the listed reactions are the reactions of major importance in this text. Most of these reactions were reviewed during the discussion of monosaccharides in Section 6.3.

10.6   Yes. Identical molecules are chemically and biologically equivalent regardless of their origin. If the vitamin C from rose hips contained contaminants, the biological activity of this preparation might differ from that of pure vitamin C (both synthetic and purified from rose hips).

10.7   See Table 10.6

10.8   True. Many reactions are catalyzed by enzymes that rely upon vitamin-derived coenzymes for their catalytic activity. In the absence of vitamins, no coenzymes are made and the coenzyme-requiring enzymes are inactive.

10.9   Vitamins, by definition, are organic compounds.

10.10   It is very difficult to control and monitor vitamin intake when using humans as experimental subjects. Moral and ethical concerns also limit the types of experiments that can be performed. In addition, one's vitamin requirements are impacted by smoking, drugs, sunlight, the intestinal flora, and other variables (see answer to Exercise 10.11). Experimental animals are of limited value, since their vitamin requirements differ from those of humans.

10.11   Age, smoking habits, drug interactions (if any), genetics, nature of your intestinal flora, amount of exposure to sunlight, gender, what other compounds are present in your diet, general health, and other variables.

10.12   True. Some vitamins, including vitamin A, vitamin D, pyridoxine and niacin, are toxic at high doses.

10.13   Hormones directly or indirectly modulate the catalytic activity of enzymes, but they are not themselves catalysts. Enzymes are, by definition, catalysts.

10.14   Hormones, in contrast to vitamins, are chemical messengers and are produced by the body. Most vitamins, in contrast to hormones, are precursors to coenzymes, and all vitamins are essential nutrients. A small number of hormones are produced from vitamins.

10.15   False. Receptors for polar hormones must be on the outer surface to be accessible to the hormones. Without a transporter, polar compounds cannot readily pass through the plasma membrane to reach internal receptors.

10.16   Hormone receptors distinguish target cells from nontarget cells.

10.17   A reaction cascade is a sequence of enzyme-catalyzed reactions in which an initial enzyme-catalyzed reaction triggers one or more secondary reactions by modulating the catalytic activities of one or more enzymes. A secondary reaction may trigger one or more additional reactions, and so on. Catalysts alter catalysts in a reaction cascade.

10.18   No. Since polar cAMP molecules are unable to pass through the plasma membrane of liver cells, cAMP in blood is unable to reach those proteins in liver cells that are targets for internally-generated cAMP.

10.19   True. Hormone receptors are protein, and the binding sites on proteins are pH sensitive. The binding of a hormone to a receptor is analogous to the attachment of a substrate to the active site on an enzyme, a process known to be affected by alterations in pH (Section 5.5).

10.20   True. Antagonist, like many competitive inhibitors, compete with natural ligands for a common attachment site. If an antagonist or inhibitor is bound, the normal ligand (the hormone or substrate) cannot bind.

10.21   Proteins, nonprotein peptides, steroids, eicosanoids, and nonpeptide amines.

10.22   NO and CO help regulate blood flow, penile erection, blood clotting, nerve action and immune response. They activate quanylate cyclases and stimulate the production of cGMP, a common second messenger.

10.23   Diacylglycerols and inositol-1,4,5-trisphosphates are produced from phosphatidylinositol-4,5-bisphosphates. Diacylglycerols dock to and activate membrane-bound, Ca^{2+}-dependent protein kinase C's. Inositol-1,4,5-trisphosphate binds to specific receptors on the endoplasmic reticulum and stimulates the release of stored Ca^{2+}.

10.24   G-protein cycling hydrolyzes GTP and consumes energy. Since the action of every kinase involves the cleavage of ATP, each step in a kinase cascades consumes energy. In the epinephrine-linked cascade, energy is also consumed during the production of cAMP from ATP.

10.25   Dopamine, epinephrine, norepinephrine and histamine are the examples identified in this chapter. Several other examples have been discovered, including serotonin.

10.26   Protein X + ATP $\rightleftarrows$ ADP + phosphorylated protein X

10.27   The binding of GTP to a G protein creates a complex that either activates or inhibits adenylate cyclase, the enzyme that catalyzes the production of cAMP. The binding of the GTP throws a molecular switch that "turns on" the adenylate cyclase-modulating ability of the G protein.

10.28   False. Phosphorylation at any site on an enzyme will impact side chain interactions at that site and tend to alter the folding of the enzyme as a whole. A change in folding will, in turn, tend to modify the size, shape and side chain interactions at the active site, that site where substrates bind and catalysis occurs.

10.29   The G-protein has a built in GTPase activity that limits the length of time it remains in the "on" position, that position which stimulates or inhibits adenylate cyclase. The hydrolysis of the GTP temporarily inactivates the G-protein.

Cyclic nucleotide phosphodiesterases catalyze the hydrolysis of cAMP to yield products with no second message activity.

Protein phosphatases catalyze the hydrolytic dephosphorylation of those proteins phosphorylated by protein kinases, and, in the process, disrupts the signaling pathway.

10.30   Since hormone and antagonist are competing for the hormone receptor, Le Chatelier's principle predicts that, at very high concentrations, the hormone will totally (from a practical standpoint) displace the antagonist. Similarly, high substrate concentration can totally displace a competitive inhibitor from the active site of an enzyme (Section 5.10).

10.31   False. An agonist competes with a hormone for its receptor. When the agonist is bound, the hormone cannot bind and vice versa. The hormone binds with equal affinity in the presence and absence of an agonist.

10.32   No. Neurotransmitter receptors are proteins, and the binding sites on proteins are normally highly specific; they will usually accommodate only a single ligand or a small group of structurally similar ligands. Since most of the neurotransmitters are structurally dissimilar, it is unlikely that the same receptor would accommodate two distinct transmitters.

10.33   True. Neurotransmitters function as ligands to open ion channels in their receptors. Once open, the influx of Na^+ alters the voltage (electrical potential) across the membrane and triggers the opening of additional ion channels (Figure 10.19).

10.34   Normal polarization is produced by the Na^+K^+ transporter which exports 3 Na^+ for every 2 K^+ imported. This makes the cytocylic side of the plasma membrane negative relative to the outside. The influx of negative Cl^- enhances the charge gradient and leads to too much (hyper) polarization.

10.35   All will bear one or more charges. Each of these substances is a carboxylic acid, an aliphatic amine, a quaternary ammonium salt or a phosphoric acid ester, four classes of compounds that exist in a charged form at pH 7. Carboxylic acids and phosphoric acid esters normally have pK_a values well below 7 and exist predominantly in a dissociated, negatively charged form at pH 7 (Section 2.7). Aliphatic amines tend to have pK_a values significantly above 7 and exist mainly in a positively charged protonated form at pH 7. Quaternary ammonium salts are permanently charged. Study Figures 10.5 and 10.20. The structure of amino acids and peptides are presented in Chapter 3, while the structure of ATP is examined in Chapter 7.

10.36   Anabolic: transcription, building muscle.

Catabolic: digestion, conversion of glycogen to glucose 1-phosphate.

10.37   True. Enzymes are "compartmentalized" within such complexes; their location and mobility is restricted since they are confined to the complexes.

10.38   A cell must be able to quickly modify its behavior in response to changes elsewhere in the body or to changes in its own needs. When hormone levels drop, it is normally desirable that the response triggered by the hormone be shut down. Short half-lives for second messengers help accomplish this.

10.39   Study Figure 10.26. Glucose from glycogen and glucose synthesized from noncarbohydrate components enters the blood. Fatty acids are converted to ketone bodies which are also released into the blood. Since both the glucose and ketone bodies can penetrate the blood-brain barrier, they help insure that the brain does not run low on fuel. Liver cells usually conserve glucose by oxidizing proteins and amino acids to satisfy their own energy needs.

10.40   Muscle cells convert glucose to glycogen when resting. That glycogen serves as a fuel reserve for the next round of muscle activity. Study Figure 10.24.

10.41   Adipocytes release fatty acids into the blood when blood glucose levels are low (Figure 10.25). These fatty acids tend to be used as a fuel for many different cells, including muscle cells. The use of this alternate fuel conserves glucose for use by brain cells. Brain cells can take up blood glucose but not blood fatty acids.

10.42   Study Table 10.11. The hormones are insulin, glucagon, epinephrine and norepinephrine. The primary target cells for insulin are hepatocytes, muscle cells and adipocytes. For glucagon, they are hepatocytes (mainly) and adipocytes. Epinephrine and norepinephrine target muscle cells, adipocytes, hepatocytes, and the pancreas.

10.43   Brain cells, in contrast to muscle cells, have virtually no stored fuel and can utilize a very limited number of alternate (other than glucose) fuels. Brain cells quickly run short on fuel and energy when blood glucose levels drop too low.

10.44   A candy bar or some other snack rich in glucose or sucrose (table sugar, a disaccharide containing one glucose and one fructose residue).

11

The Òxidation of Glucose

11.1   Muscle contraction and other movement

Active transport across membranes, including nerve action

Growth

Signal transduction and amplification

Tissue repair/turnover

Maintenance of normal body temperature

See Exhibit 8.1 and related discussions.

11.2   The energy used by the human body comes from the oxidation of dietary organic compounds. These organic compounds are derived directly or indirectly from plants. Plants produce organic compounds through photosynthesis, a process that utilizes energy form sunlight to reduce CO_2 (Introduction to Chapter 8 and Chapter 15).

11.3   Aerobic glycolysis yields a maximum of 7 net moles of ATP, 2 moles produced during glycolysis itself and up to 5 moles generated when the 2 moles of NADH from glycolysis shuttle electrons into a mitochondrion and through the respiratory chain. Anaerobic glycolysis generates a net of 2 moles of ATP. During anaerobic glycolysis, the NADH generated during the glyceraldehyde-3-phosphate dehydrogenase-catalyzed reaction is consumed during the lactate dehydrogenase-catalyzed reaction (Figure 11.3).

11.4   Oxidation—the loss of electrons; often associated with the loss of hydrogen or the gain of oxygen.

Reduction—the gain of electrons; often associated with the gain of hydrogen or the loss of oxygen.

1,2-ethanediol, ethanal and ethanoic acid are more highly oxidized than ethanol; they all contain more oxygen and/or less hydrogen. If uncomfortable with the concept of oxidation and reduction, study Section 8.7.

11.5   A "kinase" is a transferase that catalyzes the transfer of a phosphoryl group from ATP to a second substrate. Phosphoglycerate kinase and pyruvate kinase are not exceptions to this definition. These enzymes are classified as kinases on the basis of the reversal of the reactions that they catalyze during glycolysis.

11.6   The hexokinase-catalyzed reaction ($\Delta G_p \approx -33$ kJ/mole) and the phosphofructokinase-1-catalyzed reaction ($\Delta G_p \approx -22$ kJ/mole) are furthest from equilibrium under physiological conditions. The more negative the ΔG for a reaction, the further it will go toward completion and the further it is from equilibrium (Section 8.2).

Under standard biological conditions, the pyruvate kinase-catalyzed reaction ($\Delta G^{\circ\prime} = -32$ kJ/mole) and the phosphoglycerate kinase-catalyzed reaction ($\Delta G^{\circ\prime} = -19$ kJ/mole) are the furthest from equilibrium.

11.7   Pyruvate is generated during the pyruvate kinase-catalyzed reaction.

ATP is generated during both the phosphoglycerate kinase-catalyzed reaction and the pyruvate kinase-catalyzed reaction. However, the ATP generated by the phosphoglycerate kinase-catalyzed reaction simply replaces the ATP used during earlier steps in glycolysis. The pyruvate kinase-linked ATP accounts for the net yield of ATP.

NADH + H+ are produced during the glyceraldehyde-3-phosphate dehydrogenase-catalyzed reaction.

H_2O is a product of the enolase-catalyzed reaction.

11.8   True. The drop in pH is due to an accumulation of lactic acid as lactate dehydrogenase catalyzes the reduction of pyruvate by NADH. The NAD+ formed allows glycolysis to continue in the absence of O_2.

11.9   Changes in pH alter the extent of dissociation of weak acid functional groups within polypeptides. These alterations modify those interactions that maintain the secondary, tertiary and quaternary structures of proteins. The resultant changes in conformation alter the active sites of enzymes and their catalytic capabilities (Section 5.5).

11.10   If ATP inhibited hexokinase, glycogen synthesis and the pentose phosphate pathway would tend to shut down under those conditions (high energy charge) where a cell most benefits from these metabolic processes. When the energy charge of a muscle cell is high, it needs to restock its fuel reserves (in the form of glycogen) and to use the abundant energy for additional anabolism (requires reducing power from the pentose phosphate pathway).

11.11   No. In contrast to the phosphoglucoisomerase-catalyzed reaction (interconverts glucose 6-phosphate and fructose 6-phosphate), the phosphofructokinase-1-catalyzed reaction (generates fructose 1,6-bisphosphate) is virtually irreversible under physiological conditions.

11.12   True. Glycolysis is more rapid under anaerobic conditions since the concentration of positive allosteric effectors (including ADP and AMP) increases and the concentration of negative effectors (including ATP, NADH, and citrate) decreases. When the respiratory chain is inactivate due to the lack of O_2, ATP tends to be consumed more rapidly than it is formed and glycolysis speeds up. This helps a cell avoid a potentially disastrous ATP shortage. Under normal aerobic conditions, most of the ATP produced by a human cell is generated in the respiratory chain.

11.13   False. Allosteric sites and active sites are normally separate, structurally-distinct sites. Effectors often differ markedly from substrates. Phosphofructokinase-1 (ATP and glucose 6-phosphate are substrates), for example, is allosterically inhibited by citrate, and pyruvate kinase (ADP and PEP are substrates) is allosterically activated by fructose 1,6-bisphosphate.

11.14   There is no net production of ATP. In the absence of a functional triose phosphate isomerase, only one of the two 3-carbon fragments produced from glucose can be converted to pyruvate during glycolysis. This yields 2 ATPs that replace the 2 ATPs consumed in the early stage of glycolysis. Hence, the net production of ATP is zero.

         When coupled to the respiratory chain, the single NADH produced during glycolysis in the mutant cell can lead to the formation of approximately 2.5 net ATP.

11.15 NAD^+ and lipoic acid. NAD^+ is the primary oxidizing agent. Lipoic acid removes 2 electrons from each pyruvate that enters the pyruvate dehydrogenase complex. Although FAD functions as an oxidizing agent in the pyruvate dehydrogenase complex and the citric acid cycle, it does not directly oxidize glucose. In the pyruvate dehydrogenase complex, FAD returns lipoic acid to its oxidized form after the lipoic acid has oxidized glucose-derived pyruvate. In the citric acid cycle, FAD helps regenerate oxaloacetate after an acetyl group from glucose has been oxidized. NADH and $FADH_2$ deliver the 24 electrons from the complete oxidation of glucose to the respiratory chain.

11.16   The reactions catalyzed by the pyruvate dehydrogenase complex, isocitrate dehydrogenase, and the α-ketoglutarate dehydrogenase complex account for most of the CO_2 generated within the human body. The latter two enzymes are involved in the oxidation of fatty acids as well as carbohydrates.

11.17   The isomerization reactions are those catalyzed by phosphoglucoisomerase, triose phosphate isomerase, and phosphoglycerate mutase. Mutases are a subclass of isomerases.

11.18   Oxidoreductases: glyceraldehyde-3-phosphate dehydrogenase

Lyases: aldolase and enolase

Transferases: the four kinases (hexokinase, phosphofructokinase-1, phosphoglycerate kinase, and pyruvate kinase)

Table 5.1 describes the 6 major classes of enzymes.

11.19   The hydrolysis of enoyl phosphates releases considerably more energy than the hydrolysis of phosphate anhydrides. When the cleavage of an enoyl phosphate is coupled to the formation of a phosphate anhydride, considerable energy (−31.4 kJ/mole in the case of pyruvate kinase-catalyzed reaction) is released in the process. Reread Section 8.6 if you are unable to follow this answer.

11.20   The abnormal genes encode one of the polypeptides used to construct lactate dehydrogenase (Section 5.11). While an inactive lactate dehydrogenase has no impact on aerobic glycolysis, it prevents anaerobic glycolysis by preventing the regeneration of NAD^+.

11.21   Allosteric enzymes have distinct binding sites for effectors, and effector binding modulates enzyme activity (Section 5.12). Allosteric enzymes are usually oligomeric enzymes that exhibit sigmoidal, rather than hyperbolic, curves when initial velocity (y-axis) is plotted against substrate concentration (x-axis).

11.22   Fructose 1,6-bisphosphate feeds forward in glycolysis to activate pyruvate kinase. This keeps the pyruvate kinase-catalyzed reaction from becoming rate-limiting when the main floodgate for glycolysis (phosphofructokinase-1) is wide open and fructose 1,6-bisphosphate concentrations are high. Under these conditions, it is normally desirable for glycolysis to operate at a maximum rate.

CoASH and NAD^+ both feed forward and activate the pyruvate dehydrogenase complex. This makes sense because a high concentration of either CoASH or NAD^+ signals a low energy charge and a need to move more pyruvate into the citric acid cycle. On the other hand, when the energy charge in a cell is high, most of the NAD^+ has been converted to NADH and the majority of the CoASH is attached to acetyl groups, citryl groups or other activated fuels.

Glucose 6-phosphate feeds back to inhibit hexokinase. When glucose 6-phosphate concentrations are high, a cell is not using much glucose for glycolysis, glycogen synthesis or the pentose phosphate pathway. Under these conditions, it is often best (from the standpoint of the body as a whole) to export the glucose. At any rate, it makes little sense to activate glucose when activated glucose is not in demand.

ATP feeds back to inhibit phosphofructokinase-1, pyruvate kinase, the pyruvate dehydrogenase complex, citrate synthase, and isocitrate dehydrogenase. Thus, ATP inhibits its own production. It is wasteful to continue to burn fuel to make more ATP when an abundance of ATP is already present.

Acetyl-SCoA feeds back to inhibit the pyruvate dehydrogenase complex. A high acetyl-SCoA concentration represents a high energy charge, since the activated acetate is so readily oxidized to yield ATP. When the energy charge is high, the furnace needs to be turned down.

Citrate feeds back to inhibit phosphofructokinase-1 and, in the process, glycolysis. Abundant citrate signals a high energy charge and adequate intermediates for anabolism. Under these conditions, glucose should normally be conserved.

NADH feeds back to inhibit the pyruvate dehydrogenase complex, citrate synthase, isocitrate dehydrogenase, and the α-ketoglutarate dehydrogenase complex. Like ATP, high concentrations of NADH signal a high energy charge. Under such conditions, the rate of fuel oxidation should be reduced.

In vitro, succinyl-SCoA feeds back to inhibit citrate synthase and the α-ketoglutarate dehydrogenase complex. This inhibition may not be significant *in vivo*. The molecular logic behind this feedback regulation is under investigation.

11.23   Yes. ATP is a negative modulator of phosphofructokinase-1 and pyruvate kinase. A drop in ATP reduces the degree of ATP-linked inhibition. Since a drop in ATP is normally associated with an increase in ADP or AMP or both, the lifting of ATP-linked inhibition is usually coupled to a direct activation of glycolysis by ADP or AMP or both.

11.24   True. Mitochondria are the power houses of cells. A cell that uses more energy needs more power houses.

11.25   The cytosolic enzymes are those involved in glycolysis and those that participate in the shuttling of electrons from NADH into mitochondria. The remainder of the enzymes are confined to mitochondria.

11.26   Six. A separate binding site is most likely required for each effector. Positive effectors bind at different sites than negative effectors. Each of the three positive effectors is structurally so unique that a separate binding site is required for each. The same is true for the three negative effectors. Binding sites on proteins are usually highly selective.

11.27   NAD⁺, FAD, the oxidized form of lipoic acid, and Coenzyme Q (ubiquinone).

11.28

Coenzyme	Vitamin Precursor	Role Within Pyruvate Dehydrogenase Complex
TPP	Thiamine (B_1)	Carries "active" acetaldehyde.
Lipoic Acid	None	Oxidizes acetaldehyde to acetic acid. Carries the acetyl group.
CoASH	Pantothenic Acid	Carries acetyl group.
FAD	Riboflavin (B_2)	Returns reduced lipoic acid to its oxidized form.
NAD⁺	Niacin	Returns $FADH_2$ to FAD

11.29   Protein kinases—substrates are ATP and proteins

Protein phosphatases—substrates are H_2O and phosphorylated proteins.

See Section 5.13 for a detailed discussion of enzyme regulation through phosphorylation.

11.30   The redox reactions are the four steps in which NAD⁺ or FAD are reactants. These steps are catalyzed by isocitrate dehydrogenase, the α-ketoglutarate dehydrogenase complex, the succinate dehydrogenase complex, and malate dehydrogenase.

11.31   True. The membranes that surround organelles often prevent a compound produced within an organelle from mixing with an outside pool of the compound. Retained compounds are "channeled" into reactions within the organelle.

11.32   ATP inhibits both pathways, while ADP activates both.

Shared effectors make common sense, because glycolysis and the citric acid cycle share common functions. When a cell needs a large amount of ATP (leads to high ADP and low ATP concentrations), both glycolysis and the citric acid cycle should be activated. When a cell has an adequate supply of immediately useful energy as signaled by high ATP and low ADP concentrations, the furnace (includes both glycolysis and the citric acid cycle) should be turned down.

11.33   The flavins and quinones are the carriers capable of transporting 2 electrons.

Iron-sulfur clusters, cytochromes, and copper atoms can accommodate only one electron at a time.

11.34   1 ATP. Roughly 1.5 ATP can be produced when electrons from $FADH_2$ flow through both Complex III and Complex IV. Since Complex IV pumps twice as many protons, it accounts for two-thirds of the ATP ($1.5 \times 2/3 = 1$).

11.35   True. In the absence of O_2, the terminal electron acceptor within the respiratory chain, electrons "back up" along the chain and all of the electron carriers accumulate in their reduced forms. Electron flow comes to a standstill. In the presence of O_2, the reduced forms of the carriers are continuously converted back to oxidized forms as electrons are passed on down the respiratory chain to O_2.

11.36   Cysteine.

11.37   An iron atom is the electron acceptor in both cytochromes and iron-sulfur clusters. Iron has two stable oxidation state, Fe^{3+} and Fe^{2+}. An iron atom oscillates between these two states as electrons flow through cytochromes and iron-sulfur clusters.

11.38   Structurally, a and c cytochromes differ in their heme groups and polypeptide chains (see Figure 11.19). Functionally, they differ in reduction potential and, in some cases, mobility.

11.39   Iron is involved in multiple processes that are crucial to the oxidation of fuels. Iron plays a central role in the transport of O_2 from the lungs to other tissues (hemoglobin is inactive without iron), and it is required to assemble part of the enzymes in the citric acid cycle. Iron is also an electron carrier within the respiratory chain.

11.40   ATP synthase is directly responsible for ATP production. The energy for ATP production comes directly from an electrochemical gradient (H^+ gradient) created by proton pumps imbedded in the inner mitochondrial membrane.

11.41   27%. The production of 25 mol of ATP traps 775 kJ (25 mol ATP $\times$ 31 kJ/mol ATP). Since the complete oxidation of 1 mol of glucose releases 2823 kJ, the efficiency of ATP production is 27% [(775/2823) $\times$ 100]. Thus, twenty seven percent of the energy released during the oxidation is converted to a biologically useful form.

11.42   a) pH gradient decreases

b) ATP production decreases

An uncoupler allows H^+ to reenter the matrix without passing through ATP synthase. This diminishes the pH gradient. Since the pH gradient provides the energy for ATP synthesis, its reduction leads to a drop in ATP synthesis.

11.43   Cyanide leads to an increase in the rate of glycolysis and a decrease in the rate of cycling by the citric acid cycle. By blocking the flow of electrons through the respiratory chain, cyanide leads to an accumulation of NADH and a drop in NAD^+ concentration in mitochondria. This slows the citric acid cycle since NAD^+ is a required reactant and NADH is an allosteric inhibitor. The inhibition of the citric acid cycle and respiratory chain creates a low energy charge that leads to the stimulation of glycolysis. When the respiratory chain is blocked, the NAD^+ needed for glycolysis is produced by the lactate dehydrogenase-catalyzed reaction.

11.44   The hexokinase and phosphofructokinase-1-catalyzed reactions use energy provided by ATP. An H^+ gradient provides the energy for the synthesis of ATP by ATP synthase. The H^+ gradient, itself, is created and maintained with energy from the redox reactions along the respiratory chain. Although not mentioned in this chapter, the transporter that moves pyruvate (from glycolysis) into mitochondria employs energy from an electrochemical gradient (H^+ gradient).

12

Additional Topics in Catabolism

12.1   Brain cells, in contrast to muscle cells, have little stored fuel. In addition, the blood-brain barrier prevents many potential fuel molecules from entering the brain. Under normal conditions, brain cells rely mainly on blood-supplied glucose to meet their fuel and energy needs.

12.2   A candy bar or other food item rich in sucrose or glucose. Sucrose is rapidly digested to yield glucose plus fructose. The absorption of glucose from the intestine can quickly boost blood glucose levels.

12.3   Glucose from liver glycogen and gluconeogenesis (synthesis of glucose from noncarbohydrate precursors) is released into the blood when blood sugar levels are low. Glucose from blood is incorporated into liver glycogen when blood sugar levels are too high.

12.4   In muscle:

Epinephrine is principal regulatory hormone.

Glycogen phosphorylase b and phosphorylase b kinase are both regulatory enzymes. AMP (+), ATP (−), and Ca^{2+} (+) are effectors.

Mobilized glucose is oxidized within the glycogen-containing cell.

In liver:

Glucagon is the principal regulatory hormone.

Glycogen phosphorylase a is a regulatory enzyme. Glucose (–) is an effector.

Mobilized glucose tends to be exported for oxidation by extrahepatic tissues, including brain.

12.5   Liver cells produce a glucose-6-phosphatase not found in muscle cells. This enzyme catalyzes the conversion of glucose 6-phosphate (cannot pass through the plasma membrane and it has no transporters) to free glucose (can pass through transporters in the plasma membrane). *To a very large extent, the enzymes within a cell determine what reactions occur within that cell.*

12.6   Muscle glycogen is an immediately-available, rapidly-mobilizable fuel that can be oxidized to provide energy for muscle contraction. With more stored glycogen, a muscle cell can repeatedly contract for a longer period of time without having to rely on an external fuel supply.

12.7   Specific protein kinases are activated by the binding of hormones to membrane receptors. These enzymes participate in reaction cascades that ultimately lead to the phosphorylation and activation of glycogen phosphorylase, the enzyme responsible for the phosphorolysis of glycogen. In muscle, phosphorylase b kinase is allosterically activated by Ca^{2+}.

12.8   The binding of glucagon to a membrane receptor triggers the production of cAMP. Once produced, the cAMP binds to and, in the process, activates a protein kinase that initiates a protein kinase cascade. That cascade ultimately leads to the phosphorylation and activation of glycogen phosphorylase. Glycogen phosphorylase, along with debranching enzyme, is directly responsible for the mobilization of the glucose in glycogen.

12.9

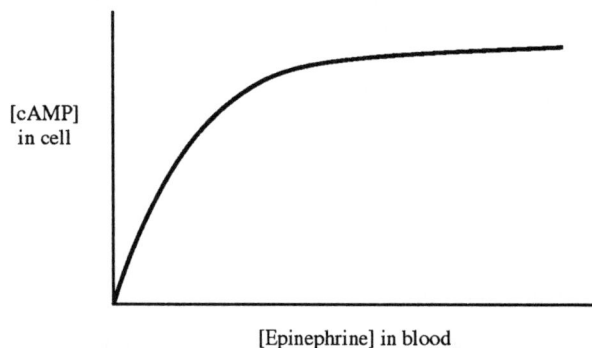

A hyperbolic curve is expected because high epinephrine concentrations can saturate receptors in the same manner that high substrate concentrations saturate active sites.

12.10

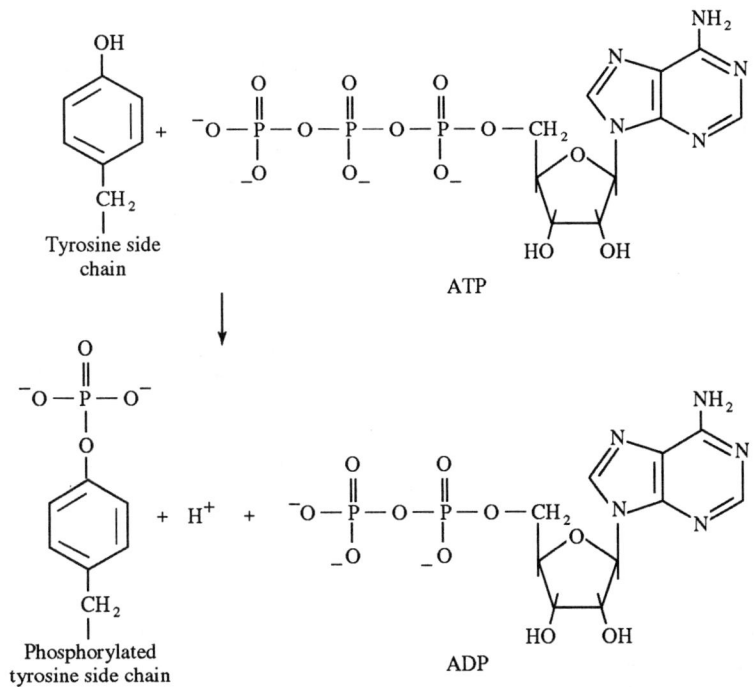

12.11

Hydrolysis

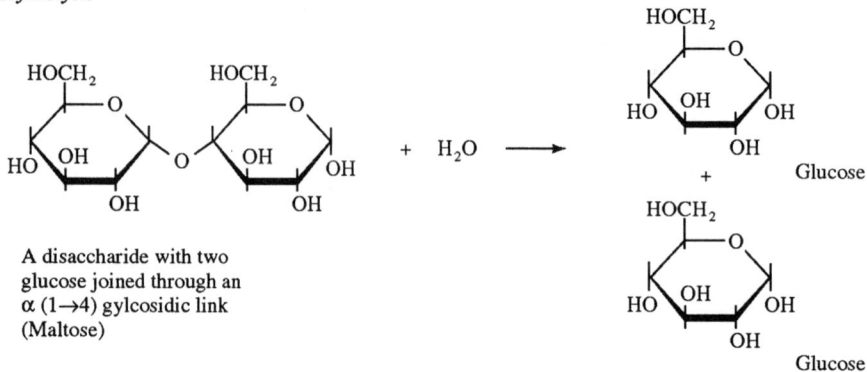

A disaccharide with two
glucose joined through an
α (1→4) gylcosidic link
(Maltose)

Glucose

+

Glucose

Phosphorolysis

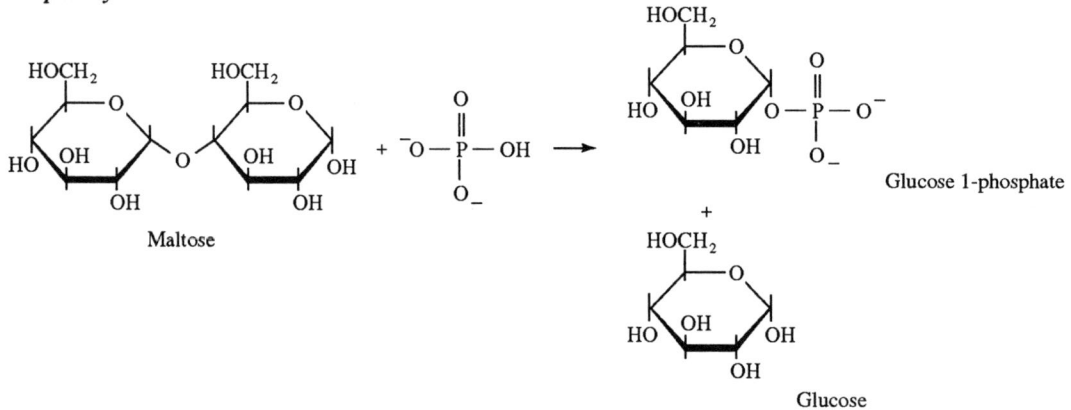

Maltose

Glucose 1-phosphate

+

Glucose

Glycogen phosphorylase catalyzes the phosphorolysis
of glycogen in liver and muscle cells.

12.12   The entry of glucose into muscle cells is stimulated by insulin. When functional insulin is
in short supply, as is common in insulin-dependent diabetics, muscle cells are unable to
import much glucose, and they are forced to rely on other fuels, primarily fat.

12.13   Insulin-dependent diabetes:

Reduced production of normal insulin

Normal target cell responsiveness to insulin

Ketosis is common

Non-insulin-dependent diabetes:

Insulin levels normal

Reduced responsiveness of target cells to insulin

Ketosis is uncommon

12.14   A diabetic coma can be due to ketoacidosis or hypoglycemia (due to over-administration of insulin). Administration of insulin could lead to death if a coma were due to hypoglycemia, since insulin enhances hypoglycemia by stimulating the uptake of glucose from blood.

12.15   False. No ATP is produced within the pentose phosphate pathway, and NADPH, in contrast to NADH, does not usually feed electrons into the respiratory chain. NADPH is normally used directly for anabolism.

12.16   The pentose phosphate pathway is primarily employed to provide pentoses and reducing power for anabolism. Glycolysis provides intermediates and ATP for anabolism. Glycolysis, when coupled to the respiratory chain, yields more ATP than when operating in isolation.

12.17   High NADPH levels signal that there is already an adequate supply of reducing power and pentoses for anabolism. Under these conditions, glucose should be stored or used for some purpose other than the pentose phosphate pathway-linked production of more NADPH and pentoses.

12.18   Ethanol production enables organisms to produce net ATP through glycolysis in the absence of oxygen. Under these conditions, the respiratory chain is nonfunctional and the ATP from glycolysis may be essential to survival. The conversion of pyruvate to ethanol continually regenerates the NAD^+ needed to support glycolysis. Ethanol, itself, is of no direct benefit to the cells capable of alcoholic fermentation.

12.19   acetaldehyde + NADH + $H^+ \rightleftarrows$ ethanol + NAD^+

The explanation lies in Le Chatelier's principle. In liver cells under aerobic conditions, the ratio of NADH to NAD^+ tends to be low (NADH is rapidly converted to NAD^+ as it transfers electrons to the respiratory chain), while that ratio is high in fermenting yeast (NAD^+ is rapidly converted to NADH when the rate of glycolysis increases in response to a low energy charge generated by anaerobic conditions). A high NADH to NAD^+ ratio favors the forward reaction (as presented immediately above) while a low ratio favors the reverse reaction.

12.20   The acetaldehyde from the pyruvate decarboxylase-catalyzed decarboxylation is reduced to ethanol. The acetaldehyde from the pyruvate dehydrogenase complex-catalyzed decarboxylation is attached to TPP to generate an "active aldehyde" (Figure 11.11). The "active aldehyde" is subsequently oxidized to an acetyl group that is transferred to CoASH. Although not discussed in this chapter, the acetaldehyde produced during the pyruvate decarboxylase-catalyzed reaction is also attached to a TTP prior to its release from the enzyme complex.

12.21   Two net ATP. Two ATP are consumed early in glycolysis, while 4 ATP are generated as two glyceraldehyde 3-phosphates are catabolized to pyruvate (Figure 11.2). Since there is no oxygen around under conditions where ethanol is formed, the NADH from glycolysis cannot be used to make ATP. This NADH is converted to NAD^+ as acetaldehyde is reduced to ethanol.

12.22   Adrenaline helps mobilize fuel for muscle action. By binding to receptors on target cells, it leads to the release of glucose 1-phosphate from muscle glycogen and the release of fatty acids from adipocytes (Table 10.11). When a person is in danger, muscles need fuel for fight or flight.

13

Additional Topics in Catabolism

13.1   Glycolysis, with its coupled production of ATP, is kept operational under anaerobic conditions by the conversion of pyruvate to lactate in the cytosol, a reaction that regenerates the NAD^+ needed for glycolysis. Fatty acid oxidation shuts down completely in the absence of O_2 since all of the mitochondrial pathways involved (β-oxidation, citric acid cycle and respiratory chain) grind to a halt. The oxidized forms of the coenzymes involved (NAD^+ and FAD) are depleted as the reduced forms of the coenzymes (NADH and $FADH_2$) pile up in mitochondria. The NAD^+ produced during the formation of lactate in the cytosol cannot enter mitochondria. Even if β-oxidation remained operational under anaerobic conditions, this pathway would be unable to generate ATP since β-oxidation, in contrast to glycolysis, entails no substrate-level phosphorylation.

13.2   The cascade activates lipases that catalyze the hydrolysis of fats.

13.3   Different cells contain distinct cAMP-dependent kinases or distinct combinations of kinases. In the presence of cAMP, these distinct cAMP-dependent kinases initiate unique kinase cascades by catalyzing the phosphorylation (activation) of different kinases or different combinations of kinases.

13.4 β-oxidation, citric acid cycle and respiratory chain.

13.5   "Acyl-" is a generic term that refers to the group remaining after the —OH is removed from the carboxyl group within any carboxylic acid. "Acetyl-" and "acetoacetyl-" are specific acyl groups.

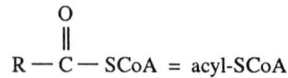

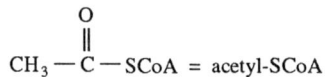

$$R - \overset{\overset{\displaystyle O}{\|}}{C} - SCoA = \text{acyl-SCoA}$$

$$CH_3 - \overset{\overset{\displaystyle O}{\|}}{C} - SCoA = \text{acetyl-SCoA}$$

$$CH_3 - \overset{\overset{\displaystyle O}{\|}}{C} - CH_2 - \overset{\overset{\displaystyle O}{\|}}{C} - SCoA = \text{acetoacetyl-SCoA}$$

13.6

$$\overset{⑤}{\downarrow} \quad \overset{④}{\downarrow} \quad \overset{③}{\downarrow} \quad \overset{②}{\downarrow} \quad \overset{①}{\downarrow}$$
$$C-C-C-C-C-C-C-C-C-C-C-\overset{\overset{\displaystyle O}{\|}}{C}-OH$$

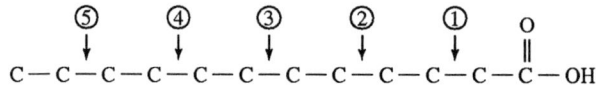

5 Rounds of β-oxidation are required with one NADH and one FADH$_2$ produced during each round.

Net ATP production

Formation of lauryl-SCoA	– 1	ATP
*Hydrolysis of PP$_i$ from formation of lauryl-SCoA	– 1	ATP
**5 NADH from 5 rounds β-oxidation	+12.5	ATP
**5 FADH$_2$ from 5 rounds β-oxidation	+ 7.5	ATP
†6 GTP from 6 acetyl-SCoA put through citric acid cycle	+ 6	ATP
**18 NADH from 6 acetyl-SCoA put through citric acid cycle	+45	ATP
**6 FADH$_2$ from 6 acetyl-SCoA put through citric acid cycle	+ 9	ATP
Net	+78	ATP

*Hydrolysis of PP$_i$ is equivalent to the hydrolysis of an anhydride bond in ATP.
**Assumes that electrons from the reduced coenzymes are passed to O$_2$ through the respiratory chain. The listed ATPs are generated within the respiratory chain.
†One GTP (an ATP equivalent) is produced for each acetyl group oxidized within the citric acid cycle.

$$\text{% of net ATP generated within the respiratory chain} = \frac{72}{78} \times 100 = 92\%$$

The 92% assumes that 2 of the 74 ATP produced within the respiratory chain are used to replace the 2 ATP used during the formation of lauryl-SCoA.

13.7   This feedback inhibition blocks the production of more acetyl-SCoA when there is already an adequate supply. The overproduction of acetyl-SCoA can be dangerous, since high concentrations of this metabolite favor the formation of ketone bodies. At high concentrations, ketone bodies lead to ketoacidosis, coma and even death.

13.8

Hexanoic Acid

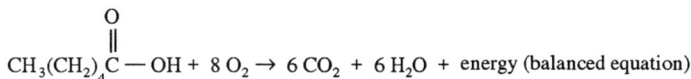

$$CH_3(CH_2)_4\overset{\overset{\displaystyle O}{\|}}{C}-OH + 8\,O_2 \rightarrow 6\,CO_2 + 6\,H_2O + \text{energy (balanced equation)}$$

8 O_2 consumed during total oxidation

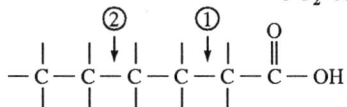

Two rounds of β-oxidation generate 3 acetyl-SCoA, 2 NADH and 2 FADH$_2$. The oxidation of the 3 acetyl groups in the citric acid cycle yields 3 GTP, 9 NADH and 3 FADH$_2$. Since 2 ATP equivalents are consumed to activate the fatty acid (form an acyl-SCoA), the maximum net yield of ATP is 36. This assumes that each NADH yields 2.5 ATP and each FADH$_2$ yields 1.5 ATP (within the respiratory chain).

molecular weight = 116 atomic mass units (u)

$$\frac{\text{Net moles ATP per}}{\text{gram oxidized}} = \frac{36\text{ mol ATP}}{1\text{ mol hexanoic acid}} \times \frac{1\text{ mol hexanoic acid}}{116\text{ g hexanoic acid}} \times 1\text{ g hexanoic acid} = 0.31\text{ mol ATP}$$

Glucose

$$C_6H_{12}O_6 + 6\,O_2 \rightarrow 6\,CO_2 + 6\,H_2O + \text{energy (balanced equation)}$$

6 O_2 consumed during total oxidation

molecular weight = 180 u

The complete oxidation of 1 mole of glucose yields a maximum of 32 net moles of ATP (Table 11.4).

$$\frac{\text{Net moles ATP per}}{\text{gram oxidized}} = \frac{32\text{ mol ATP}}{1\text{ mol glucose}} \times \frac{1\text{ mol glucose}}{180\text{ g glucose}} \times 1\text{ g glucose} = 0.18\text{ mol ATP}$$

Comparison

The oxidation of 1 g hexanoic acid yields 0.13 more moles of ATP (0.31–0.18).

This is a 72% increase over the amount of ATP produced from 1 g glucose.

The difference would be even greater if a longer-chained, saturated fatty acid had been used for the calculations.

13.9   Yes. Several of the 20 common protein amino acids can be directly catabolized to oxaloacetate or to other central metabolites (including pyruvate) that can readily be converted to oxaloacetate (Figure 13.20). An increase in oxaloacetate concentration lowers the concentration of acetyl-SCoA by increasing its rate of entry into the citric acid cycle. As acetyl-SCoA levels drop, the rate of ketone body production also decreases.

13.10

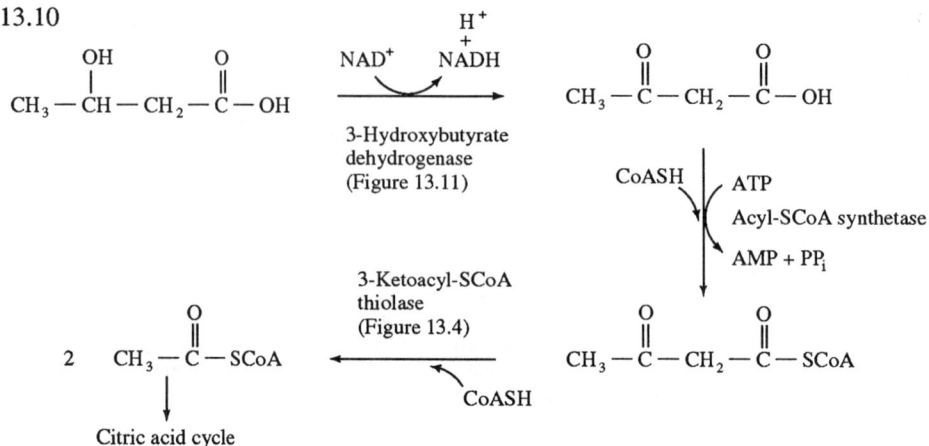

Net ATP Production		
NADH from β-hydroxybutyrate dehydrogenese reaction		+ 2.5 ATP
Acyl-SCoA synthetase reaction with coupled hydrolysis of PP$_i$ (see Table 13.1)		− 2 ATP
6 NADH from 2 acetyl-SCoA that enter the citric acid cycle		+15 ATP
2 FADH$_2$ from 2 acetyl-SCoA that enter the citric acid cycle		+ 3 ATP
2 GTP from 2 acetyl-SCoA that enter the citric acid cycle		+ 2 ATP
	Net	20.5 ATP

Alternatively, 3-hydroxybutyrate could be attached to CoASH before it was oxidized to oxaloacetate. Net ATP production would be the same.

13.11   The kidneys have to excrete more $^+NH_4$ and urea. Since fats and most dietary carbohydrate contain no nitrogen, their catabolic products (CO_2 and water) are less toxic and more readily eliminated from the body. CO_2 is exhaled while water is lost through the skin, lungs and kidneys.

13.12   Hydrogen—ends up in water and urea

Carbon—ends up in CO_2

Nitrogen—ends up in $^+NH_4$ and urea (primarily)

Sulfur—ends up in SO_2

Oxygen—ends up in CO_2

This answer assumes that the carbon skeletons are completely oxidized during catabolism.

13.13   Reactions where $^+NH_4$ is a product:

a) Glutamate dehydrogenase-catalyzed reaction—releases $^+NH_4$ from collected amino groups so that the nitrogen in the amino group can be attached to a transport agent for transport to the liver. In the liver, most of the nitrogen from transported amino groups is incorporated into urea for excretion.

b) Glutaminase-catalyzed reaction—releases $^+NH_4$ transported from extrahepatic tissues to the liver by glutamine so the $^+NH_4$ can be incorporated into urea.

Reactions where $^+NH_4$ is a reactant:

a) Glutamine synthetase catalyzed reaction—$^+NH_4$ is attached to glutamate (glutamine is formed) for transport from extrahepatic tissues to the liver (site of urea production). $^+NH_4$ is too toxic to be released into the blood without being attached to a carrier.

b) Carbamoyl phosphate synthetase I-catalyzed reaction—CO_2, $^+NH_4$ and phosphate are coupled to form carbamoyl phosphate which enters the urea cycle. The nitrogen in the $^+NH_4$ ultimately ends up in urea.

All of these reactions are involved in controlling the toxicity of the compounds produced during the nitrogen-linked catabolism of amino acids. Collectively, the reactions help to collect protein nitrogen in the liver and to incorporate this nitrogen into urea (a relatively nontoxic compound) for excretion. Study Figures 13.17 and 13.18.

13.14   Aspartate and glutamate. The α-keto acid from each of these amino acids is a citric acid cycle intermediate.

13.15   When on a high protein diet, large amounts of amino acid nitrogen need to be catabolized and eliminated from the body. The urea cycle needs a boost in order to accomplish this. During prolonged starvation, the body runs low on carbohydrates and fats and begins to catabolize relatively large amounts of protein in order to produce the energy essential for survival. An increase in the concentration of urea cycle enzymes is required to support this catabolism.

13.16   The activation of fatty acids (formation of acyl-SCoAs) is "pulled" by PP_i hydrolysis (Table 13.1) as is the formation of argininosuccinate within the urea cycle (Figure 13.18). The hydrolysis of PP_i removes a product from the reaction mixture. If the reaction is at equilibrium before the PP_i is removed, it will undergo a net forward reaction to restore equilibrium after the PP_i is removed (Le Chatelier's principle).

13.17   Fumarate can be converted to oxaloacetate (within the citric acid cycle), a compound that can be used for gluconeogenesis (Figure 13.12). Acetoacetate is a ketone body.

13.18   Melanin can be made from dietary tyrosine even in the absence of phenylalanine. Individuals with PKU, who are unable to convert phenylalanine to tyrosine, tend to produce less melanin than normal individuals, because their dietary tyrosine is not supplemented with phenylalanine-derived tyrosine.

14

Anabolism

14.1   False. The concentration of a protein is determined by the relative rates of synthesis and degradation (assuming that degradation is the only route through which protein is lost). If the rate of synthesis is greater than the rate of degradation (regardless of how rapid the rate of degradation), protein concentration will increase. Protein concentration will remain at any initial level whenever the rate of synthesis equals the rate of degradation. A short half-life signals a rapid rate of degradation, but it provides no information about the concentration of a protein or its rate of synthesis.

14.2   No. The activity of a pool of proteins can be altered by changes in pH, effector concentrations, and other variables. The biological activity of a pool of proteins is determined by two factors: 1) protein concentration; and 2) the environment of the protein.

14.3   True. All of the 4-carbon intermediates in the citric acid cycle can be converted to oxaloacetate, the product of the first step in gluconeogenesis from pyruvate (Figure 14.3). Amino acids, lactate, and glycerol are normally the major precursors for gluconeogenesis.

14.4   The liver is the site of most gluconeogenesis. A majority of the glucose produced in the liver is exported for use by the brain and other tissues.

14.5

Enzyme	Major Class*
Pyruvate carboxylase	Ligase
PEPCK	Lyase
Enolase	Lyase
Phosphoglycerate mutase	Isomerase
Phosphoglycerate kinase	Transferase
Glyceraldehyde-3-phosphate dehydrogenase	Oxidoreductase
Triose phosphate isomerase	Isomerase
Aldolase	Lyase
Fructose-bisphosphatase	Hydrolase
Phosphoglucoisomerase	Isomerase
Glucose-6-phosphatase	Hydrolase

*To arrive at the correct answer, one must sometimes consider the reverse reaction.

14.6   Maximum life span is directly or indirectly programmed into DNA, that molecule which carries inherited information. This proposal is known as the **programmed theory** of aging.

14.7   The free energy content of reactants and products; $\Delta G^{o\prime} = -RT \ln K_{eq}{}'$ and $\Delta G_p = \Delta G^{o\prime} + RT \ln ([C]^c[D]^d/[A]^a[B]^b)$ (Section 8.2). A reaction with a large equilibrium constant (large negative $\Delta G^{o\prime}$) is irreversible (from a practical standpoint) under standard conditions. A reaction is reversible under standard conditions if $\Delta G^{o\prime}$ is near 0 and $K_{eq}{}'$ is close to 1.

14.8   True. Since 1,3-bisphosphoglycerate can transfer a phosphoryl group to ADP to form ATP (a high energy compound), 1,3-bisphosphoglycerate must itself be a high energy compound. The cleavage of one high energy compound commonly provides the thermodynamic driving force for the formation of a second high energy compound.

14.9   Under anaerobic conditions, a low energy charge [high (ADP + AMP)/ATP ratio] makes gluconeogenesis unfavorable; ATP is required for gluconeogenesis while AMP is an allosteric inhibitor. In addition, pyruvate produced by glycolysis in the cytosol tends to be converted to lactate before it can be transported into mitochondria for oxaloacetate formation. The formation of lactate also consumes the cytosolic NADH needed for gluconeogenesis. As mitochondrial acetyl-SCoA concentrations drop under anaerobic conditions, the extent to which acetyl-SCoA allosterically activates pyruvate carboxylase is also reduced.

14.10   All of the enzymes for liver gluconeogenesis from pyruvate are cytosolic enzymes with the exception of pyruvate carboxylase (mitochondria) and glucose-6-phosphatase (endoplasmic reticulum).

14.11   Liver cells contain glucose 6-phosphatase which catalyzes the hydrolysis of glucose 6-phosphate to yield free glucose. Once freed, glucose can move across the plasma membrane through glucose transporters. Muscle cells lack glucose phosphatases and possess no mechanism for transporting glucose phosphates across the plasma membrane.

14.12   The complete oxidation of glycerol involves part of glycolysis, the pyruvate dehydrogenase complex, the citric acid cycle, and the respiratory chain. The conversion of glycerol to glucose involves part of the enzymes required for gluconeogenesis from pyruvate. Both processes also involve two enzymes (glycerol kinase and glycerol phosphate dehydrogenase) that catalyze the conversion of glycerol to dihydroxyacetone phosphate, an intermediate in both glycolysis and gluconeogenesis (Figure 13.2). Effectors that stimulate glycolysis or inhibit gluconeogenesis favor oxidation over glucose synthesis. These effectors are AMP and fructose 2,6-bisphosphate.

14.13   The effectors will most likely bind to separate sites, since it is unlikely that a single binding site can accommodate each of the three structurally diverse compounds. Compounds must normally be structurally similar before they can compete with one another for binding to the same site on a protein. The anticancer drugs described in Section 14.15 support this claim. Section 5.10 (enzyme inhibitors and substrate inhibition) provides additional support.

14.14   AMP and fructose 2,6-bisphosphate favor glycolysis while citrate and ATP favor gluconeogenesis. Acetyl-SCoA also tilts the metabolic balance in favor of gluconeogenesis by activating pyruvate carboxylase (catalyzes oxaloacetate production from pyruvate) and inhibiting the pyruvate dehydrogenase complex (catalyzes the conversion of pyruvate to acetyl-SCoA and CO_2). High levels of citrate, ATP and/or acetyl-SCoA normally signal a high energy charge, and these three effectors all favor gluconeogenesis over glycolysis. When the energy charge is high, fuel should be conserved and fuel reserves should be beefed up.

14.15   A substrate cycle is a futile cycle when it provides no benefits. Cycles that consume energy and serve no benefit are, in reality, worse than futile; they are detrimental.

14.16   The Cori cycle indirectly shifts part of the burden for ATP production from muscle to liver during anaerobic exercise.

14.17   Once inside muscle, glucose becomes trapped through phosphorylation, since muscle lacks a phosphoglucose phosphatase (see Exercise 14.11). This prevents the glucose from moving from muscle to the liver.

14.18   Little, if any, lactate accumulates under aerobic conditions. Lactate is a dead end product of anaerobic glycolysis.

14.19 O_2 concentration is determined by the balance that exists between the rate of O_2 loss and consumption and the rate of O_2 delivery. Most of the O_2 consumed by a cell serves as the terminal electron acceptor for the respiratory chain. In general, the more rapidly a cell uses ATP, the more rapidly it consumes O_2 and the lower its O_2 concentration. The rate of O_2 delivery is determined by rate of blood flow, the concentration of red blood cells and hemoglobin, the respiration rate, the concentration of O_2 in inhaled air, and other variables.

14.20   The body is unable to use acetyl-SCoA from any source to build glucose, because it is not genetically programmed to produce enzymes capable of catalyzing the conversion of the acetyl group to gluconeogenic precursors. Acetyl groups entering the citric acid cycle are converted to CO_2, not to gluconeogenic precursors.

14.21   The conversion of glucose to the fatty acid components of triacylglycerols involves glycolysis (to generate pyruvate), the pyruvate dehydrogenase complex (converts pyruvate to acetyl-SCoA) and the enzymes of fatty acid biosynthesis. Two of the carbons from each glucose and part of the oxygen in glucose are lost as CO_2 within the pyruvate dehydrogenase complex. Some oxygen is also lost during glycolysis and during fatty acid biosynthesis from acetyl-SCoA. This explains why 10 grams of glucose yield less than 10 grams of fat. Triacylglycerols, however, contain more calories per gram since they are in a more reduced state than glucose. The glycerol 3-phosphate used for triacylglycerol biosynthesis is produced from glucose by reducing dihydroxyacetone phosphate, an intermediate in glycolysis.

14.22   Biotin is a B vitamin and the precursor for a coenzyme named biocytin. Most biotin-containing enzymes catalyze carboxylation reactions (Table 10.6). Pyruvate carboxylase (gluconeogenesis) and acetyl-SCoA carboxylase (fatty acid biosynthesis) are specific examples.

14.23   Ketones: reduction to alcohols; hemiacetal (once called hemiketals) and acetal (once called ketals) formation; aldol condensation; tautomerism; plus other reactions.

    Alcohols: oxidation to aldehydes, carboxylic acids or ketones; hemiacetal and acetal formation; ether formation; ester formation; reduction to alkanes (ribose to deoxyribose, for example); dehydration to alkenes; plus other reactions.

    Alkenes: addition of water; addition of halogens; reduction to alkanes; oxidation to diols and oxidative cleavage to yield ketones, carboxylic acids, and/or carbonic acid; plus other reactions.

    Thioesters: hydrolysis; transesterification; plus other reactions.

    Carboxylic acids: decarboxylation; ester formation; anhydride formation; reduction to aldehydes and alcohols; amide formation; salt formation; plus other reactions.

14.24   There are no common intermediates since CoASH serves as the acyl carrier for β-oxidation while ACPSH functions as the acyl carrier for biosynthesis. Ignoring the difference in carriers, 3-oxoacyl (β-ketoacyl) groups, trans-2,3-alkenes, and acyl groups are common intermediates. The configuration about the β-alcohol intermediate in β-oxidation is L, while the configuration about the β-alcohol intermediate during fatty acid biosynthesis is D.

14.25   The major energy-consuming steps in gluconeogenesis are those catalyzed by pyruvate carboxylase, PEPCK, and phosphoglycerate kinase. All of these are coupled to the hydrolysis of ATP or GTP. If reducing power is considered a form of energy (it can be used to generate ATP within the respiratory chain), glyceraldehyde-3-phosphate dehydrogenase also catalyzes an energy-consuming transformation.

Similarly, the major energy-consuming step in fatty acid biosynthesis is catalyzed by acetyl-SCoA carboxylase. If reducing power is considered a form of energy, the 3-oxoacyl–[ACP] reductase and enoyl-[ACP] reductase-catalyzed reactions are also energy-consuming.

14.26   Peptide (amide) bonds; the enzymes are all part of a single polypeptide chain.

14.27

Enzyme	Major Class
Acetyl-SCoA carboxylase	Ligase
[ACP] maloryltransferase	Transferase
[ACP] acetyltransferase	Transferase
3-Oxoacyl-[ACP] synthase	Transferase
3-Oxoacyl-[ACP] reductase	Oxidoreductase
3-Hydroxyacyl-[ACP] dehydratase	Lyase
Enoyl-[ACP] reductase	Oxidoreductase

14.28   The enzymes that use NADPH as a substrate are unable to use NADH. Enzymes, in general, are highly specific catalysts.

14.29   Prostaglandins and other eicosanoids.

14.30   Regular exercise leads to a hormone-linked increase in the concentration of the enzymes involved in fat catabolism and a decrease in the level of the enzymes responsible for fat production.

14.31   A combination of both. Although regulation centers around the modulation of the activity of existing enzymes, insulin does lead to long-term changes in enzyme concentrations in liver.

14.32   An organism would run the risk of running low on fuel or energy as a consequence of this futile cycle. The cycle would also tie up coenzymes that might be needed by other metabolic pathways.

14.33

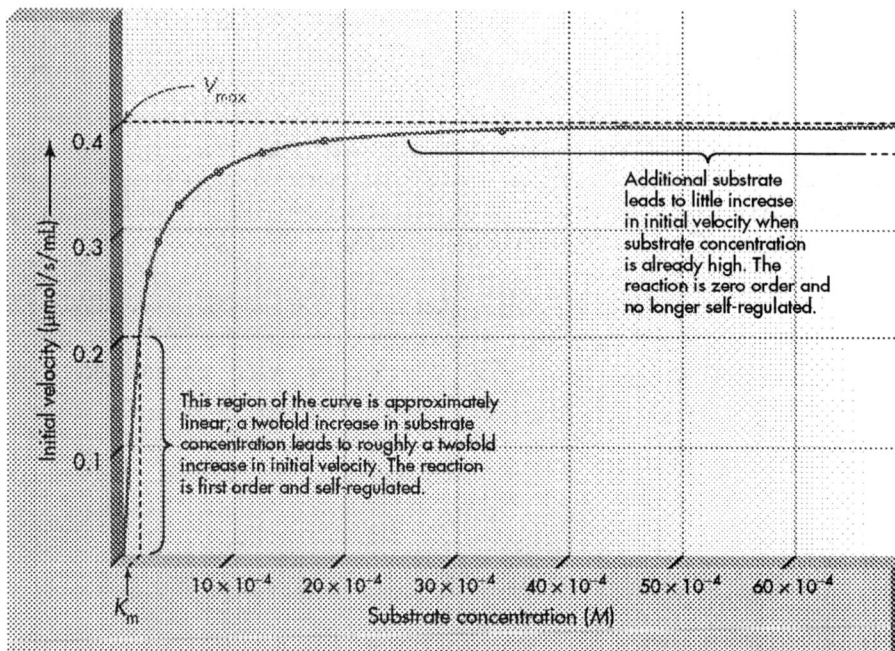

This graph reveals that substrate concentration can have a very marked impact on the rate of an enzyme-catalyzed reaction. At substrate concentrations far below K_m, a doubling of substrate concentration approximately doubles the initial velocity. The rate of an entire metabolic pathway can be altered by changing the rate of the rate-limiting step for that pathway.

14.34   4 ATP: one to produce glycerol 3-phosphate and one to form each of three acyl-SCoAs (Section 14.9).

14.35   A thioester bond is cleaved as an ester bond is formed. The $\Delta G°'$ is negative, since a thioester possesses more free energy than a oxyester (Sections 8.2 and 8.5). The $\Delta G°'$ is negative whenever the reactants possess more free energy than the products. The more negative the $\Delta G°'$, the larger the equilibrium constant for a reaction.

14.36   Most bacteria are genetically programmed to synthesize the enzymes needed to catalyze the production of the "essential" amino acids.

14.37   Nucleotide = nucleoside monophosphate = phosphate-sugar-base

Nucleoside = sugar-base

Nucleoside diphosphate = phosphate-phosphate-sugar-base

Nucleoside triphosphate = phosphate-phosphate-phosphate-sugar-base

ATP = nucleoside triphosphate containing the base adenine

AMP = nucleotide containing the base adenine

RNA = a polymer of ribonucleotides (contain the sugar ribose)

Nucleic acid = a polymer of nucleotides. Includes RNA and DNA.

14.38   PRPP (5-phosphoribosyl-1-pyrophosphate). The ribose phosphate residue in this compound is activated since it is attached to the rest of the compound through a phosphate anhydride link. PRPP is produced from ribose 5-phosphate generated within the pentose phosphate pathway.

14.39   Ribose residues must be converted to deoxyribose residues, and uracil residues to thymine residues.

14.40   DNA polymerase, the enzyme that catalyzes the joining of monomers during DNA biosynthesis, mistakes dUTP for dTTP and uses the dUMP as a substrate.

14.41   The anticancer drugs examined in this chapter target rapidly dividing cells, including cancer cells, hair cells and certain intestine cells, in which there is rapid DNA replication. Damage to rapidly dividing normal cells is a side effect of the therapy.

14.42   5-Fluorouracil inhibits thymidylate synthase, that enzyme which catalyzes the conversion of dUMP to dTMP. DeoxyTMP is one of the four essential building blocks for DNA (Section 7.4). Without TTP (the activated form of dTMP and a substrate for DNA polymerase), DNA replication is impossible.

14.43   By competing with methotrexate for the active site on dihydrofolate reductase, dihydrofolate would tend to block methotrexate binding; when dihydrofolate is bound, methotrexate cannot bind and vice versa. Methotrexate is not as effective as an anticancer agent when cancer cells contained high concentrations of dihydrofolate.

14.44

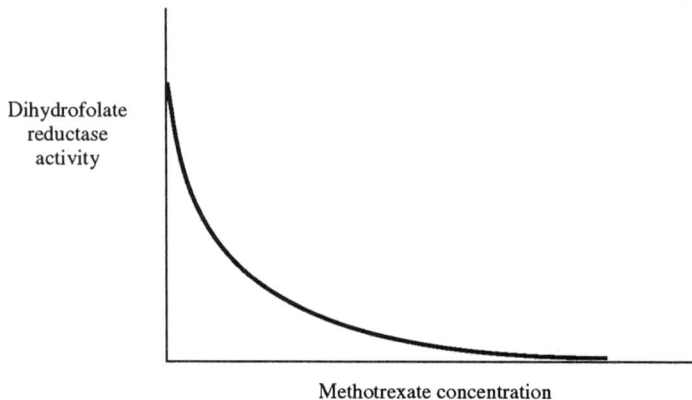

Methotrexate concentration

At high methotrexate concentrations, virtually all of the dihydrofolate
binding sites on dihydrofolate reductase are occupied by methotrexate and
the enzyme is almost totally inhibited. Since methotrexate must displace the
dihydrofolate through binding competition, an inverted hyperbolic curve is
obtained.

14.45   One can, in theory, target any of the enzymes involved in DNA replication, purine biosynthesis, pyrimidine biosynthesis or deoxynucleoside triphosphate production. These include
DNA polymerases and most of the enzymes identified in Figures 14.21 through 14.27.
However, the inhibition of some of these enzymes might also block other cellular processes and lead to severe damage to most normal cells, as well as cancer cells.

14.46   To a very large extent, the enzymes produced by an organism determine what reactions
occur in that organism. Most of the reactions occurring in the human body would not
proceed at a significant rate in the absence of enzymes (catalysts). The DNA inherited by
an organism determines what enzymes it is able to assemble.

15

Photosynthesis and
Nitrogen Fixation

15.1   True. The energy your body uses comes directly from the foods you eat. If you consume phototrophs, such as potatoes, wheat and carrots, the energy in these organisms can be traced to sunlight trapped during photosynthesis. If you consume heterotrophs, such as chickens, cows and fish, the energy in these organisms came from phototrophs further down the food chain.

15.2   Blue light is of shorter wavelength. Energy content is inversely proportional to wavelength; as wavelength increases, energy content decreases.

15.3   There are three likely reasons: a) at the surface of the earth, visible light is the most abundant form of electromagnetic radiation in sunlight (Figure 15.4); b) UV radiation is of such high energy that it would be difficult to design photosynthetic machinery that would not be damaged by the UV; and c) IR radiation is of such low energy that an enormous amount would need to be harvested, a requirement that would lead to a relatively inefficient process.

15.4   False. Some compounds undergo "allowed" transitions between energy levels that are brought about by the absorption of IR radiation rather than the absorption of visible light. For radiation to be absorbed, its energy content must correspond to an "allowed" transition between two energy levels.

Two energy levels can be close together (leads to the absorption of low energy electromagnetic radiation) or far apart (leads to the absorption of high energy electromagnetic radiation).

15.5

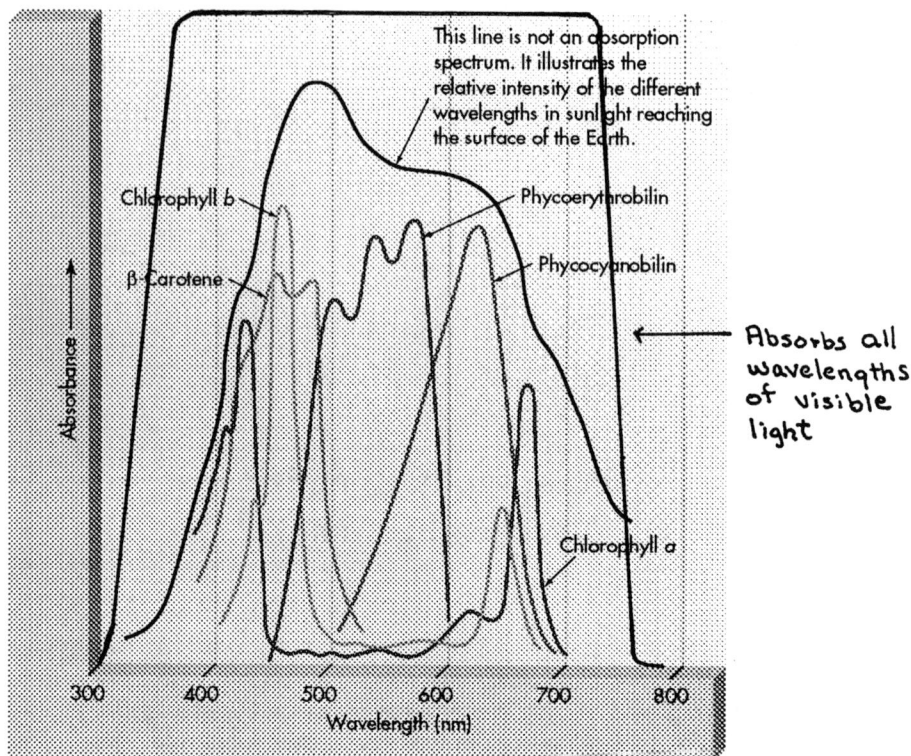

Any compound that absorbs all wavelengths of visible light appears black, because none of the visible components of sunlight are reflected back into the eye of the beholder.

15.6 Fe^{2+} is converted to Fe^{3+} when it loses a negative electron. Cu^{2+} is transformed into Cu^+ when it accepts a negative electron. Electrons flow through both iron and copper atoms as they traverse the photosynthetic electron transport system:

$$Fe^{2+} \rightleftarrows e^- + Fe^{3+}$$

$$Cu^{2+} + e^- \rightleftarrows Cu^+$$

$$Fe^{2+} + Cu^{2+} \rightleftarrows Fe^{3+} + Cu^+$$

15.7   Extensive regions of conjugation (alternating double and single bonds) primarily account for the absorption of visible light. This arrangement leads to "allowed" transitions between electronic energy levels that accommodate the absorption of visible photons.

15.8   With different absorption spectra, antenna molecules are able to harvest wavelengths of visible light that cannot be harvested by P680 and P700. This leads to more efficient use of available energy, since a larger fraction of the available energy is harvested.

15.9   A good antenna molecule must transfer the energy it absorbs to neighboring photosynthetic pigments before the energy is lost through fluorescence, heat generation, or ionization. Many compounds that absorb visible light do not satisfy this requirement.

15.10   An increase in photosynthetic pigment concentration will lead to more efficient collection of the light that is available; a larger fraction of the available light energy will be harvested because absorbance tends to be roughly proportional to the concentration of absorbing molecules: $A = \varepsilon bc$.

15.11   The direct transfer does not normally occur because photosystem II (contains P680) and photosystem I are physically separated on the thylakoid membrane and $NADP^+$ is associated with photosystem I. In addition, the enzyme that catalyzes the reduction of $NADP^+$ would be unlikely to use P680* as a substrate. Enzymes and compartmentalization play central roles in determining what reactions occur within an organism. If the direct transfer did occur, a major proton pump would be bypassed and less ATP would be produced.

Note: A mutant green algae has been identified that can, under anaerobic conditions, reduce $NADP^+$, fix CO_2, and grow photoautotrophically in the absence of a functional photosystem I (see article by J. Barber listed in *Selected Readings*).

15.12   pH 4.5 corresponds to an $[H^+]$ of 3.16×10^{-5} M, while pH 8.0 corresponds to an $[H^+]$ of 1.00×10^{-8} M (pH = $-\log[H^+]$, so $[H^+]$ = antilog pH). Hence, there is a 3160-fold difference in $[H^+]$ when one compares the lumen with the stroma (3.16×10^{-5} divided by 1.00×10^{-8}).

15.13   A *manganese* cluster accepts electrons from water during the water-splitting reaction. *Iron* is a component of the iron-sulfur clusters in the iron-sulfur proteins that serve as electron carriers on both sides of photosystem I in the Z-scheme. Iron is also a component of the cytochrome bf complex. Electrons flow through some of the iron atoms during the electron transport process. Plastocyanin is the *copper*-containing protein that directly delivers electrons to photosystem I. The copper is reduced and then reoxidized as an electron flows through this compound. *Magnesium* sits at the center of the chlorin ring in most chlorophyll molecules. Although magnesium is structurally and functionally important, electrons do not flow directly through magnesium atoms during the electron transport process.

15.14   Low ratio. Since cyclic transfer leads to the generation of ATP, the cyclic process should and does speed up when more ATP is needed (as signaled by a low ATP/NADPH ratio). The proper ATP/NADPH ratio is critical for optimum carbon fixation.

15.15   Similarities: a) many of the electron carriers are similar; b) the flow of electrons powers proton pumps; c) both systems contain both soluble (mobile) and membrane-bound carriers; d) both systems are coupled to an ATP synthase and ATP production.

Differences: a) electrons flow from water to $NADP^+$ in chloroplasts and from NADH and $FADH_2$ to O_2 in mitochondria (O_2 is produced in chloroplasts and consumed in mitochondria); b) chloroplast electron transfer requires light; c) the specific electron carriers differ.

15.16   Since phototrophs sometimes find themselves in the dark, they must be able to acquire energy in the absence of light. Energy is required continuously for survival. In some phototrophs, many cells are never exposed to light (think big trees). These cells rely totally upon oxidative phosphorylation for energy. The overall balance between photophosphorylation and oxidative phosphorylation normally favors photophosphorylation within the plant as a whole.

15.17   Humans and other heterotrophs rely upon the O_2 and organic compounds produced by phototrophs during photosynthesis. Phototrophs need the CO_2 and H_2O produced by humans and other heterotrophs. Hence, the carbon and oxygen cycles.

15.18   The removal of 2, rather than 1, glyceraldehyde 3-phosphates would rapidly deplete the concentration of ribulose 1,5-bisphosphate and other cycle participants, and the carbon fixation cycle would grind to a halt. If more than one glyceraldehyde 3-phosphate is removed, there is not enough of this metabolite left to totally replace the ribulose 1,5-bisphosphate that initiated the cycle. Five of the six glyceraldehyde 3-phosphates produced during one cycle must remain in the cycle if participant concentrations are to remain constant.

15.19

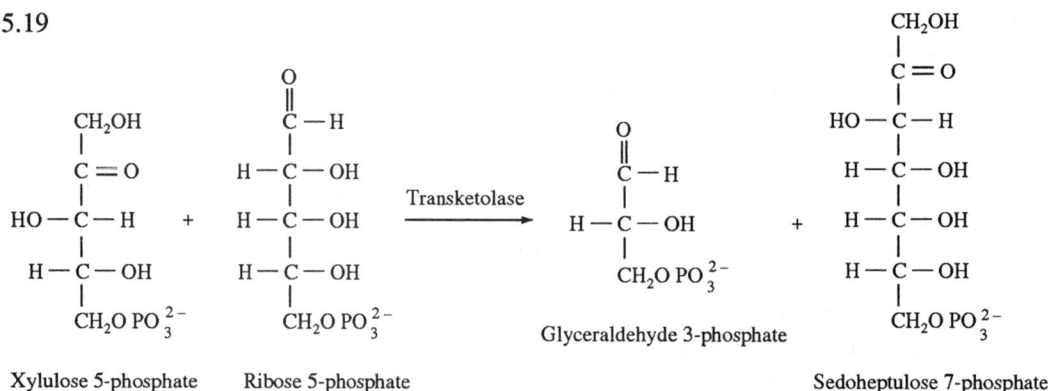

Xylulose 5-phosphate   Ribose 5-phosphate   Sedoheptulose 7-phosphate

This is the reverse of the transketolase-catalyzed reaction that generates ribose 5-phosphate and xylulose 5-phosphate during ribulose 1,5-bisphosphate regeneration (Figure 15.21). It is also a key step in the nonoxidative phase of the pentose phosphate pathway (Figure 12.10).

15.20   The irreversible reactions are those catalyzed by fructose-bisphosphatase, phosphoribulokinase, and sedoheptulose-1,7-bisphosphatase. The phosphatase-catalyzed reactions are irreversible because they involve the hydrolysis of a relatively high energy phosphate ester to yield products of significantly lower energy. The phosphoribulokinase-catalyzed reaction is irreversible since it is coupled to ATP hydrolysis. At constant temperature, the reversibility of a reaction is determined by the free energy difference between reactants and products (Section 8.2).

15.21   Yes. The carbon fixation reactions do not require light; they require ATP and NADPH.

15.22   Specific enzymes catalyze the production and breakdown of 2-carboxyarabinitol 1-phosphate and fructose 2,6-bisphosphate. There are also enzymes that catalyze the reversal of the reductive activation of several disulfide-containing enzymes. In general, it is safe to assume that a separate enzyme exists for each reaction that occurs in an organism.

15.23   Since 2-carboxyarabinitol 1-phosphate is charged, it is unable to pass through cellular membranes to reach rubisco, its target enzyme. 2-carboxyarabinitol 1-phosphate must be synthesized within chloroplasts to function as a competitive inhibitor of rubisco because rubisco is confined to chloroplasts and no 2-carboxyarabinitol 1-phosphate transporters exist.

15.24   No. 2-carboxyarabinitol 1-phosphate does not act stoichiometrically. Its binding to rubisco must be reversible to allow rubisco to be quickly reactivated following the reexposure of a plant to light. In some plants, rubisco activase catalyzes the dissociation of the noncovalently bound inhibitor. The larger the binding constant for a competitive inhibitor, the more closely the inhibition approaches a stoichiometric relationship. By definition, however, a competitive inhibitor can be displaced from an enzyme by high enough concentrations of substrate.

15.25   The light reactions activate the carbon fixation reactions by: providing ATP and NADPH; creating pH and Mg^{2+} gradients that activate specific carbon fixation cycle enzymes; and providing electrons for the thioredoxin-catalyzed activation of multiple carbon fixation cycle enzymes.

15.26   False. Glycolysis (coupled to the citric acid cycle and the respiratory chain) and the light reactions of photosynthesis are the two major sources of ATP and reducing power for plant cells. Normally, at least one of the pathways needs to be active at all times.

15.27   An increase in CO_2 concentration increases the rate of carbon fixation by: increasing the substrate concentration for rubisco, the enzyme that catalyzes carbon fixation; activating rubisco through carbamylation of a lysine side chain; and inhibiting photorespiration through substrate inhibition.

15.28   Directly, the net change is zero. Indirectly, the net change is -1; the PP_i produced during the ADP-glucose pyrophosphorylase-catalyzed reaction is hydrolyzed to pull the reaction. In general, when PP_i is a product, it is immediately hydrolyzed.

15.29   Yes. Glyceraldehyde 3-phosphate is an intermediate in glycolysis. Glyceraldehyde 3-phosphate from photosynthesis can be transported to the cytosol where it has access to the enzymes capable of catalyzing its complete oxidation, a process involving part of glycolysis, the citric acid cycle, and the respiratory chain.

Glyceraldehyde 3-phosphate from photosynthesis is sometimes transported to glycolytic enzymes in the cytosol and oxidized to 3-phosphoglycerate with the associated production of ATP and NADH. The 3-phosphoglycerate is transported into chloroplasts where it is reconverted to glyceraldehyde 3-phosphate with the use of ATP and NADPH from the light reactions. The glyceraldehyde 3-phosphate can be transported back to the cytosol, and the cycle continues. This metabolic cycle allows ATP and reducing power from the light reactions in chloroplasts to be, indirectly, transferred to the cytosol.

15.30   No. A single active site can bind either O_2 (leads to photorespiration) or CO_2 (leads to carbon fixation), but not both simultaneously. CO_2 and O_2 compete for the same binding site.

15.31   True. The O_2 released during photosynthesis in chloroplasts leads to an increase in O_2 concentration and, hence, an increase in photorespiration. The binding of O_2 to rubisco during photorespiration inhibits CO_2 binding and the carbon fixation stage of photosynthesis. This is not usually a serious problem, because the O_2 rapidly diffuses away from the site of carbon fixation.

15.32 C_4 carbon fixation and C_3 carbon fixation occur in separate types of cells. This allows CO_2 to be temporarily fixed through the catalytic activity of an enzyme that is not subject to O_2 inhibition. The fixed CO_2 is delivered to the C_3 system in a separate compartment (cell) where there is relatively little O_2 and where high local concentrations of CO_2 can be generated. Note that CO_2 does not directly enter bundle-sheath cells; it is released from malate that enters these cells.

15.33

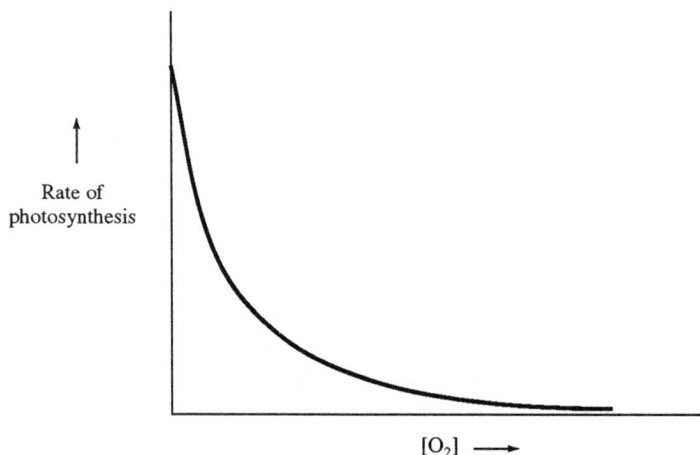

The curve is an (inverted) hyperbola. O_2 competes with CO_2 for active sites on rubisco molecules. At very high O_2 concentrations, the rate of photosynthesis will approach zero as O_2 competes virtually all of the rubisco away from CO_2.

15.34   Mesophyll cells do not produce rubisco.

15.35   No. Nocturnal inhibitor is a competitive inhibitor of rubisco. Since mesophyll cells produce no rubisco, there is no reason for them to synthesize a compound designed to regulate this enzyme.

15.36   Sixteen photons are required. Two photons are needed to move a single electron to ferredoxin through photosystems II and I, and a total of eight electrons are needed to reduce N_2 during the nitrogenase-catalyzed reaction.

15.37   False. Only certain bacteria are able to fix nitrogen, and bacteria do not contain organelles.

15.38   Yes; this is a theoretical possibility. Photorespiration consumes O_2.

15.39 NH_3 is a base; at physiological pH, it quickly reacts with H^+ to form NH_4^+, a nonvolatile ion.

16

Nucleic Acids

16.1	RNA	DNA
	Contains ribose	Contains deoxyribose
	Contains A, U, G and C	Contains A, T, G and C
	Single stranded	Double stranded
	Contains base paired and loop regions	All bases paired; two strands totally complementary

16.2

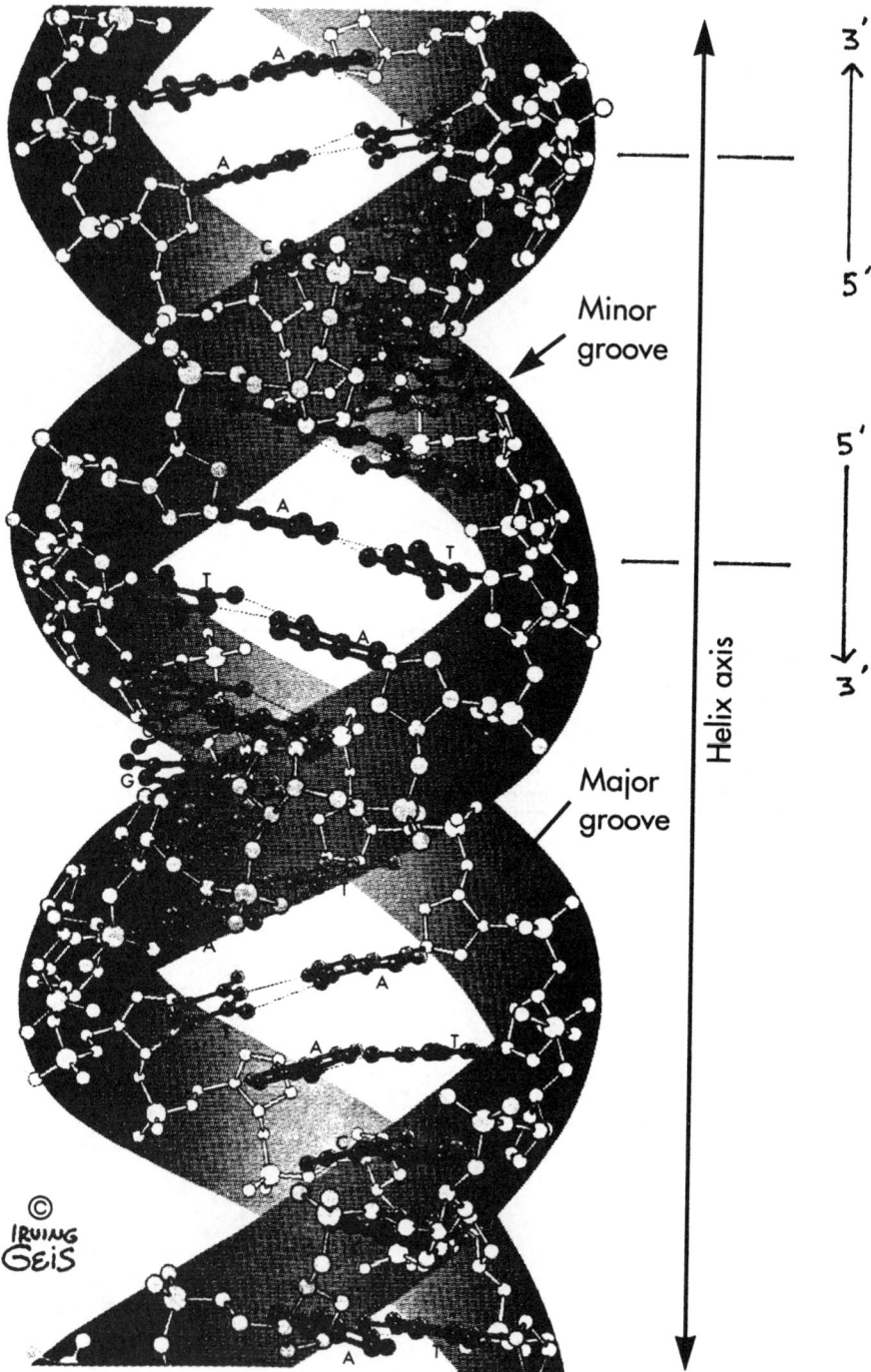

Minor groove

Major groove

Helix axis

3′
5′
5′
3′

© IRVING GEIS

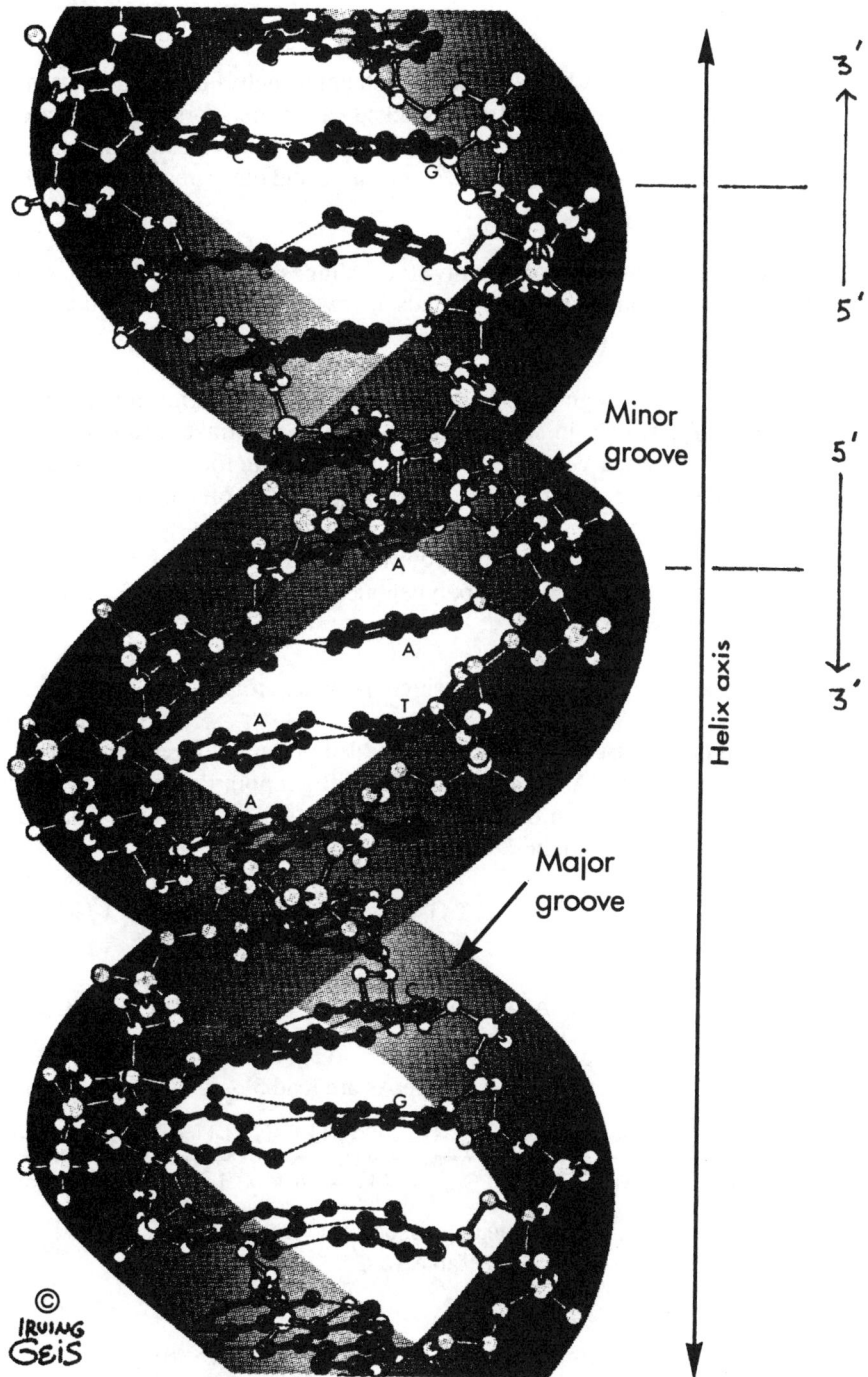

The directionality of a strand can be determined readily by identifying the 3' and 5' carbons in a clearly visible sugar residue. The 3' carbon is on that side of the sugar closest to the 3' end.

16.3   See Table 16.2.

16.4   All three forms of duplex DNA are held together through H bonding between paired bases and hydrophobic interactions between stacked base pairs. Normally, there are no covalent bonds between the two polymer strands. If covalent bonds existed, the two strands could not be reversibly separated with moderate heating and other treatments that do not rupture covalent bonds.

16.5   a) False. Such interconversions involve changes in the manner in which the two noncovalently-bonded polymer strands are twisted together; individual strands and all covalent bonds remain intact.

   b) True. A gene is a segment of DNA that carries information for the production of a specific RNA which, in some instances, carries information for the production of a specific polypeptide. Since DNA serves as a template for RNA production, the assembly of a novel mRNA normally requires the existence of a unique gene.

   c) True. In general, a unique gene is required for the assembly of each novel polypeptide. However, not all genes code for polypeptides; some genes code for rRNAs, tRNAs and other nonmessenger RNAs.

   d) False. Messenger RNAs contain a single polymer strand.

16.6   The amount of Z-DNA will increase. The added protein will remove part of the Z-DNA from the Z-DNA:B-DNA equilibrium. This will temporarily destroy the equilibrium. A spontaneous net conversion of B-DNA to Z-DNA will occur as the system moves to re-store equilibrium (Le Châtelier's Principle).

16.7   a)   5'd A-A-T-G-T-T-C-T-G-T-A-A-A-G-C-C-G-C-G-C-T-G-T-C-G-T-A 3'

↓ Transcription

3' U-U-A-C-A-A-G-A-C-A-U-U-U-C-G-G-C-G-C-G-A-C-A-G-C-A-U 5'

rewrite in a 5' to 3' direction
since mRNAs are read 5' → 3'

5' U-A-C-G-A-C-A-G-C-G-C-G-G-C-U-U-U-A-C-A-G-A-A-C-A-U-U 3'

N-terminal   Tyr - Asp - Ser - Ala - Ala - Leu - Gln - Asn - Ile   C-term

amino acid sequence read from
Table 16.6

b) 5'd T-T-G-C-A-G-C-G-T-A-G-C-T-A-G-T-A-C-G-T-A-G-C-C 3'

↓ Transcription

3' A-A-C-G-U-C-G-C-A-U-C-G-A-U-C-A-U-G-C-A-U-C-G-G 5'

rewrite in a 5' to 3' direction
since mRNAs are read 5' → 3'

5' G-G-C-U-A-C-G-U-A-C-U-A-G-C-U-A-C-G-C-U-G-C-A-A 3'

N-terminal   Gly - Tyr - Val - Leu - Ala - Thr - Leu - Gln   C-term

amino acid sequence read from
Table 16.6

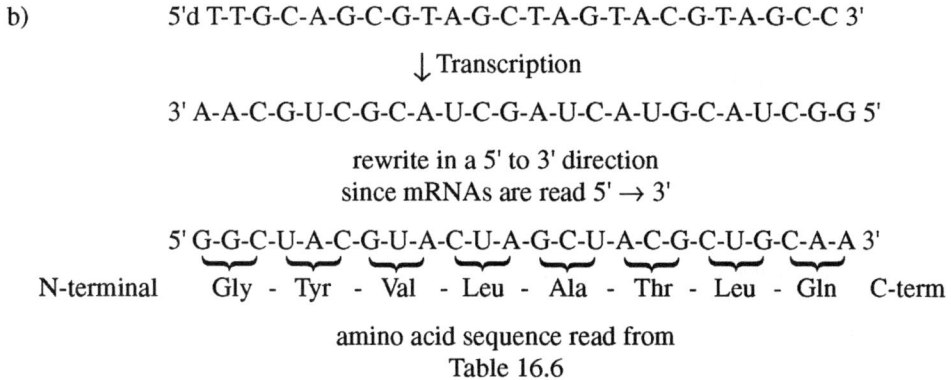

16.8   The linking number is 11.1 (100 divided by 9); the two strands of DNA wrap around each other once for each turn of the double helix.

16.9   Negative supercoiling tends to convert a right-handed helix (of the type found in B-DNA) to a left-handed helix (of the type found in Z-DNA); negative supercoiling tends to unwind a right-handed helix and to change its handedness. This can best be visualized by twisting a piece of string that is constructed from multiple smaller strands.

16.10 1.5×10^7 nucleosomes. Each nucleosome contains approximately 201 base pairs (around 146 in the core particle plus close to 55 in the linker sequence). 1×10^9 divided by 201 equals 1.5×10^7 (to two significant figures).

16.11   Proteolytic enzymes (proteinases). Since chromatin consists of both DNA and proteins (primarily histones), one can disrupt its structure by destroying either or both of these components. However, the core proteins tend to be resistant to enzyme-catalyzed hydrolysis because they are wrapped with a DNA shield. Similarly, core DNA within an intact core particle is a poor substrate for most nucleases because it is tightly associated with histones that tend to block nuclear interactions with DNA.

16.12   Most genes encode mRNAs that contain 300 to 3000 nucleotide residues (Section 16.5). Since a nucleosome contains around 200 nucleotide residues, a typical gene must encompass multiple nucleosomes.

In humans and many other organisms, a large fraction of the total DNA lies outside of genes. Such DNA also resides in nucleosomes, another reason the number of nucleosomes is not equal to the number of genes.

16.13   Genome
↓
Chromosomes
↓
Chromatin
↓
Genes
↓
Nucleosomes
↓
Core Particles

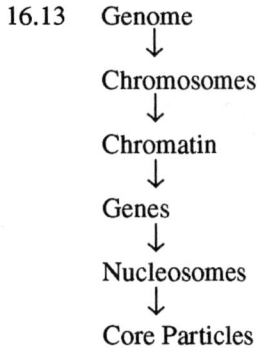

16.14   11. Two histone H2A, two histone H2B, two histone H3, two histone H4, one histone H1 and the two complementary strands of the original duplex DNA molecule.

16.15

	Substrates	Catalytic Activity	Major International Class
a) Nucleases	RNA and/or DNA, and H_2O	Hydrolysis of phosphate ester bonds	Hydrolases
b) Proteinases	Proteins, Polypeptides, and H_2O	Hydrolysis of peptide bonds	Hydrolases
c) Topoisomerases	DNA	Cleavage of DNA followed by passage and rejoining	Isomerases
d) Aminoacyl-tRNA synthetases	Amino Acids, tRNAs and ATP	The joining of amino acids to tRNAs with the coupled cleavage of ATP	Ligases

16.16   a)   5' C-U-G-A-C-C-A-A-A-A-A-C-U-G-G-C-G-A 3'

N-terminal   Leu - Thr - Lys - Asn - Trp - Arg   C-terminal

The answer is "read" from the genetic code table (Table 16.6). All mRNA are read in a 5' to 3' direction.

b)   5' G-U-U-A-C-G-C-A-C-C-C-U-C-G-G-C-G-A-U-G-A-C-C-C-A-U-G 3'

N-terminal   Val - Thr - His - Pro - Arg - Arg ←—┤——C-terminal

Termination Codon—
reading stops here

c)   5' U-C-G-U-U-U-C-A-A-C-G-C-G-C-A-G-G-U-C-C-U-A-U-U 3'

N-terminal   Ser - Phe - Gln - Arg - Ala - Gly - Pro - Ile   C-terminal

d)   5' U-U-A-U-U-G-C-U-U-C-U-C-C-U-A-C-U-G 3'

N-terminal   Leu - Leu - Leu - Leu - Leu - Leu C-terminal

16.17   a) tRNA   b) mRNA

c) tRNA, mRNA and rRNA   d) all RNAs

e) mRNA   f) tRNA

g) all RNAs   h) tRNA

i) rRNA   j) all RNAs

k) mRNA   l) tRNA

m) mRNA   n) all RNAs

o) all RNAs   p) tRNA

q) None. Messenger RNAs contain   r) tRNA
   information for the production of
   proteins, but they are not proteins.   s) mRNA

16.18   Brain cells and liver cells differ because each cell type encompasses some unique chemical reactions. Since each reaction tends to be catalyzed by a separate enzyme, both brain
cells and liver cells must contain tissue-specific enzymes. Tissue-specific mRNAs must
also exist, because distinctive enzymes are encoded in distinctive mRNAs. Although brain
cells and liver cells both contain some tissue-specific enzymes, many reactions and enzymes are common to both cell types.

16.19   A rapidly growing cell must produce many different mRNAs in order to synthesize the
enormous number of unique enzymes required to bring about those reactions that support
growth. Dormant or inactive cells assemble fewer mRNAs and enzymes since relatively
few reactions are occurring within these cells.

16.20   5' A A G 3'   Anticodon

         :: :: :::

         :: :: :::   Complementary

     3' U U C 5' ≡ 5' C U U 3'   Codon

                Leucine

     5' A C A 3'   Anticodon

         :: ::: ::

         :: ::: ::   Complementary

     3' U G U 5' ≡ 5' U G U 3'   Codon

                Cysteine

     5' G G C 3'   Anticodon

         ::: ::: :::

         ::: ::: :::   Complementary

     3' C C G 5' ≡ 5' G C C 3'   Codon

                Alanine

     5' U G U 3'   Anticodon

                      Complementary

     3' A C A 5' ≡ 5' A C A 3'   Codon

                Theonine

In each case, a tRNA will carry that amino acid specified by the codon which is complementary to its anticodon. Codons are read in a 5' to 3' direction and are antiparallel to anticodons. The amino acid specified by each codon is recorded in Table 16.6.

16.21   The minimum length for the mRNA is over 1113 nucleotide residues. There must be 1110 nucleotide residues (3 × 370) in the translated region of the mRNA, since it takes 3 separate nucleotide residues to code for each amino acid. To be translated properly, an mRNA must also contain a leader and a tailing sequence with the first three nucleotide residues in the tailing sequence constituting a termination codon. Many mRNAs also contain a PolyA tail. It is uncertain what minimum length is required for leader and tailing sequences.

16.22   Secondary structure, by definition, refers to the base paired regions within a nucleic acid. Since G and U do not pair, an RNA containing only G and U would be unable to form base paired regions.

16.23   Two mRNAs would be required to produce a single IgG molecule, one mRNA that encodes the heavy chain and a second mRNA that encodes the light chain. In eukaryotes, each mRNA usually carries information for the synthesis of a single polypeptide chain.

Only two mRNAs would be required to assemble 10 identical IgG molecules, since each mRNA can be repeatedly translated to produce many copies of the polypeptide that it encodes.

16.24   No. One can be confident that the two mRNAs do not encode the same polypeptide, because complementary codons do not code for the same amino acid. If unconvinced, write the abbreviation for a segment of an mRNA and a separate abbreviation for its complement. Use Table 16.6 to translate each segment in a 5' to 3' direction. Translation is always in a 5' to 3' direction

16.25

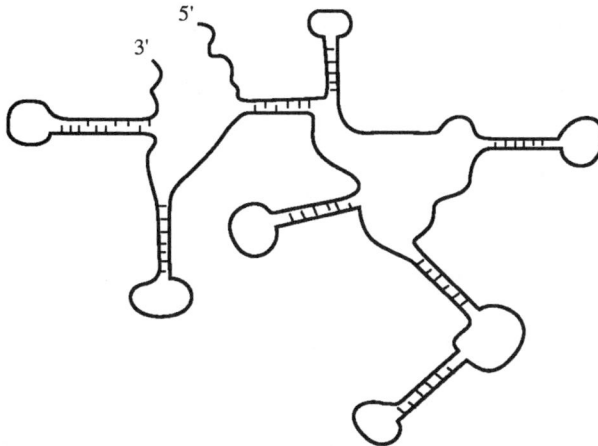

The number of possible answers is only limited by one's imagination.

16.26   UUAGCUUUUCAAUCUCCU—contains 9 U's, 3 A's, 5 C's, and 1 G
CUGGCGUUCCAGAGCCCG—contains 3 U's, 2 A's, 7 C's, and 6 G's

There are many possible answers.

16.27   An mRNA is only required when there is a need for its polypeptide product (that polypeptide that it encodes). Cells require different polypeptides at different times. Since a cell does not have much storage space and unused molecules tend to get in the way of functional ones, a molecule tends to be destroyed when it is no longer needed and resynthesized when, and if, it is once again required. The continual breakdown of existing mRNAs also helps prevent the accumulation of damaged molecules over time. Furthermore, the removal of unneeded mRNAs insures that a gene is not expressed at an inappropriate time.

16.28   This is the same as asking "how many different gene products does a rat liver cytoplasmic ribosome contain?", since there tends to be one gene for each gene product. A good answer is 86 (4 rRNAs + 82 polypeptides). You would predict that each RNA and polypeptide in the ribosome is encoded in a separate gene. In reality, some of the rRNAs are cut out of the same pre-rRNA precursor which can be considered the product of a single gene.

16.29   Denatured tRNAs will not fit the substrate binding site on aminoacyl-tRNA synthetases. Restated, a tRNA must have a native conformation to be recognized by its synthetase.

16.30   The observation indicates that the components of the 50S subunit are capable of spontaneous assembly; no chaperones or other outside agents are required. However, one cannot conclude that chaperones or other "facilitating" substances are excluded from the *in vivo* assembly process.

16.31   Ribosomal RNAs account for close to 80% (by weight) of the total RNA within a typical cell. Messenger RNAs and tRNAs make up roughly 3 to 5% and 10 to 15%, respectively. There are minor classes of RNAs, in addition to rRNAs, mRNAs and tRNAs. Some of these, including snRNAs and gRNAs, are examined in subsequent chapters.

16.32   a) 3' UAACGUGUACUACGUUAUCGU 5'

b) 3' CCCGCCCGACGCAGUGCGCAU 5'

c) 3' UUGAAAUUGAUUUAUAUAAAA 5'

For fragments to hybridize, they must be complementary and antiparallel.

16.33   b. Melting point (the temperature at which a duplex DNA molecule is half denatured) increases as G + C content increases, because there are three H bonds within each G-C pair and only two within each A-T pair. The more tightly two strands are held together, the higher the temperature required to separate them.

16.34

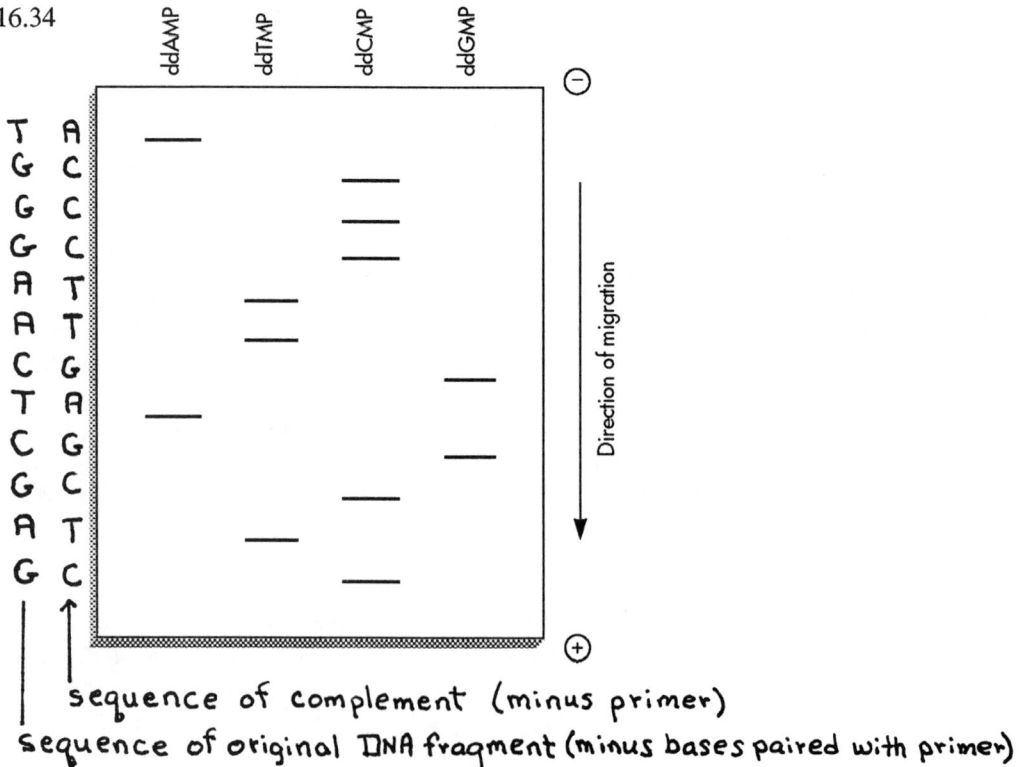

sequence of complement (minus primer)
sequence of original DNA fragment (minus bases paired with primer)

17

Nucleic Acid Biosynthesis

17.1   a) Enhancer: a segment of eukaryotic duplex DNA that, when bound to protein transcription factors, helps regulate promoter utilization.

b) RNA polymerase (transcriptase): a protein enzyme whose catalytic activity is directly responsible for the assembly of primary transcripts.

c) Antiterminator: a protein which allows transcriptases to read through some terminators.

d) Terminator: a segment of duplex DNA that, when encountered during transcription, helps trigger the termination of transcription.

e) Sigma factor: a protein which enhances the selective attachment of prokaryotic transcriptases to promoters.

f) TATA box: a nucleic acid motif within many promoters which is involved in transcriptase binding.

g) Promoter: that region of duplex DNA where transcriptase originally binds during the initiation of transcription. It contains multiple motifs and usually contains the transcription start site.

17.2   A drop in σ factor concentration would lead to a decrease in the rate of protein production. Since sigma factor is required for optimum transcription (it enhances the selective binding of transcriptase to promoters), the rate of RNA production drops as σ factor concentration is reduced. As RNA concentrations drop, the rate of protein synthesis also drops, because most RNAs are directly involved in translation. In prokaryotes, newly assembled mRNAs tend to be immediately translated with the assistance of tRNAs and rRNAs.

17.3   A single transcriptase can, in theory, catalyze the synthesis of a primary transcript of any length; the enzyme can continuously catalyze the addition of one nucleotide at a time to a growing chain (see Figure 17.9).

A minimum of 498 anhydride links are cleaved during the synthesis of a primary transcript with 250 nucleotides. The addition of each nucleotide residue directly involves the cleavage of one anhydride link. A second anhydride link is cleaved when the pyrophosphate simultaneously produced is subsequently hydrolyzed. Thus, 2 anhydride links are cleaved for each phosphodiester link that is formed. Since an RNA with 250 nucleotides contains 249 phosphodiester links, 498 (249 × 2) anhydride bonds are cleaved during the elongation process. Additional high energy bonds may be cleaved during the initiation or termination phases of transcription.

The $\Delta G^{o\prime}$ for the hydrolysis of a phosphate anhydride bond is approximately 31 kJ/mol (7.3 kcal/mole) [see Chapter 8]. No doubt about it, considerable energy is required to construct an RNA. Anabolism, in general, is an energy-consuming process.

17.4   It would take slightly over 6 days for the mRNA concentration to drop to 10% of its initial value

The initial concentration of the mRNA (concentration at the time production is shut down) is defined as 100%. The following diagram illustrates how mRNA concentration will shift over time:

100%   (Initial Concentration)
↓   1st half-life (2 days)
50%
↓   2nd half-life (2 more days)
25%
↓   3rd half-life (2 more days)
12.5%
↓   4th half-life (2 more days)
6.25%

The mRNA concentration will drop to 10% of its initial value in slightly over 6 days (a little over 3 half-lives). A more exact answer can be obtained by utilizing the half-life equation given in Exercise 3.19 (page 110).

17.5   No. During transcription, a gene simply serves as a pattern or template; it is returned to its original state at the end of the process.

17.6   False. The poly A tail is usually added post-transcriptionally; it is not normally encoded in mRNA genes.

17.7   Since different genes and different classes of genes tend to possess distinctive promoters, a drug which blocks a unique promoter will shut down only those genes whose transcription start sites are coupled to the blocked promoters.

17.8   Prokaryotic transcriptases can attach to their promoters in the absence of accessory protein factors. In contrast, transcription factors must bind to a promoter prior to eukaryotic transcriptase binding. The interactions of eukaryotic transcriptases with promoters tend to be regulated by several transcription factors. At least some of the nucleic acid and protein motifs involved in transcriptase-promoter interactions differ in prokaryotes and eukaryotes.

17.9   3'CGUUGCAACGG5' and 5'GCAACGUUGCC3' are the two RNA fragments. The one actually produced is the one complementary to the template strand. The promoter determines which strand is the template strand.

17.10   Trimming, splicing, capping, tailing and minor base formation are involved in the processing of most eukaryotic mRNA precursors (Table 17.3). Most mRNA precursors are never edited. We would reasonably predict that splicing, capping, tailing and minor base formation require energy, while trimming, which involves hydrolysis, does not. In general, energy is released when substances are broken down (catabolism) but it is required to build compounds (anabolism) [Section 10.12]. One of our predictions is inaccurate; most (but not all) introns are removed without the coupled hydrolysis of ATP or other high energy compounds.

17.11   a) **snRNPs:** ribonucleoproteins that catalyze the splicing of some primary transcripts.

        b) **hnRNAs:** nuclear RNAs that include the primary transcripts for eukaryotic mRNAs.

        c) **Exons:** the segments of mRNA precursors that are retained during splicing.

        d) **RNases:** protein enzymes that catalyze the trimming of the 3' end of primary transcripts.

        e) **Methylases:** protein enzymes that catalyze some capping events and the formation of some minor bases.

17.12   a) **gRNAs:** base pair with those segments of RNA precursors that are to be edited; guide the editing process and may be involved in catalysis.

b) **snRNAs:** found within snRNPs; base pair with intron-exon junctions during some splicing events; guide and may help catalyze splicing.

c) **hnRNAs:** contain primary transcripts for mRNAs.

d) **tRNA precursors:** are converted to tRNAs.

e) **rRNA precursors:** are converted to rRNAs.

f) **RNA primers:** involved in the initiation of the synthesis of new DNA strands.

g) **The RNA in RNase P:** a ribozyme which catalyzes the trimming of the 5' end of tRNA precursors.

h) **Telomerase RNA:** serves as a template for the extension of parent DNA strands during the replication of the ends of chromosomes; participates in the catalytic extension of the parent DNA strands.

17.13   When compared with "normal" primary transcripts, primary transcripts from cDNA lack introns and some 3' nucleotides. Complementary DNA (cDNA) is prepared from mature mRNAs which lack the introns and the extra 3' bases usually found in "normal" primary transcripts.

17.14   The loops represent those segments of the sense strand that encode introns. Since the mature mRNA contains only exons (introns have been removed), it can only hybridize with those segments of the DNA template which encode (are complementary to) exons. The exon encoding segments are separated in the template strand by intron encoding regions which loop out as the exons are brought together along the mature mRNA during hybridization.

17.15

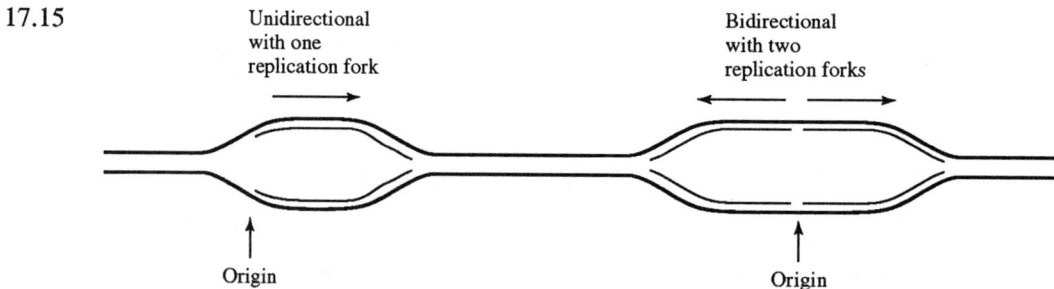

17.16   a) **helicase:** catalyzes the ATP-dependent unwinding of DNA.

b) **DNA gyrase:** a topoisomerase which relieves the torsional stress created by unwinding; it catalyzes the generation of negative supercoils in the duplex DNA ahead of the site of strand separation.

c) **SSB (single strand binding protein):** coats the separated parent strands to protect them from nucleases and to keep them from reannealing.

d) **Primase or other RNA polymerase:** catalyzes the assembly of RNA primers.

e) **Replicase:** a DNA dependent DNA polymerase that catalyzes the stepwise addition of nucleotide residues during the elongation of daughter strands. Replicase binding is considered part of the initiation stage.

17.17   DNA polymerases: a) Use deoxyribonucleoside triphosphates, rather than ribonucleoside triphosphates, as substrates and assemble DNAs rather than RNAs; b) require a primer; and c) most can proofread.

17.18   a) Proofreading

b) Removal of RNA primers

17.19   a) Helicase-catalyzed unwinding

b) DNA gyrase-catalyzed supercoiling

c) Replicase binding

d) Primer synthesis

e) Addition of nucleotide residues to daughter strands

f) Filling the gaps left by the removal of RNA primers

g) The DNA ligase-catalyzed joining ("sealing") of processed Okazaki fragments

h) Replacement of mismatched nucleotide residues during proofreading

i) Telomerase action (when applicable)

j) Mismatch repair (if considered part of replication)

k) Termination and nucleosome disruption may also be energy-dependent processes. DNA replication, like transcription, consumes an enormous amount of energy.

17.20   a) Unwinding the parent duplex

b) SSB binding to unwound DNA

c) Continuous synthesis of the leading strand of daughter DNA

 d) Discontinuous synthesis of the lagging strand of daughter DNA

 e) Maturation of the lagging strand: primers are removed from Okazaki fragments, the resultant gaps are filled, and the processed Okazaki fragments are joined.

 f) Proofreading

17.21   Transcription is a continuous process where one nucleotide reside at a time is added to the growing polymer chain. The primary transcript is complete when the last nucleotide residue is added. In contrast, the lagging strand of DNA is synthesized discontinuously with each synthesized fragment (Okazaki fragment) containing an RNA primer. The primers must be removed, the resultant gaps filled, and the processed fragments sealed, a sequence of events known as maturation.

17.22   DNA replicases contain: a) a 5' to 3' DNA dependent DNA polymerase activity that catalyzes the stepwise addition of deoxyribonucleotide residues during chain elongation; and b) a 3' to 5' exonuclease activity that catalyzes the hydrolytic removal of mismatched nucleotide residues during proofreading.

17.23   a) **Primase:** d [RNA polymerase], f [transferase], h [protein], i [enzyme]

 b) **Primer:** j [nucleic acid fragment]

 c) **Replicase:** a [exonuclease], c [DNA polymerase], f [transferase], g [hydrolase], h [protein], i [enzyme]

 d) **Okazaki fragment:** j [nucleic acid fragment]

 e) **Replicon:** j [nucleic acid fragment]

 f) **Transcriptase:** d [RNA polymerase], f [transferase], h [protein], i [enzyme]

 g) **Reverse transcriptase:** a [exonuclease], c [DNA polymerase], f [transferase], g [hydrolase], h [protein], i [enzyme]

17.24   Half-life: the time required for one half of the molecules in a sample of a compound to be degraded, inactivated or eliminated.

 Okazaki fragments have a short half-life because maturation is a rapid and continuous process. Okazaki fragments are processed and sealed together almost as rapidly as they are assembled.

17.25   a) Replicases facilitate proper matching (base pairing) of deoxyribonucleoside triphosphates with template bases

b) The primers produced during the initiation of polymerization (an error prone process) are replaced under high fidelity conditions

c) Proofreading tends to replace mismatched nucleotide resides

d) Mismatch repair systems tend to correct those mismatches which are not eliminated during proofreading

17.26   Since errors in DNA replication lead to changes in the sequence of DNA, they tend to be passed from cell to cell and, sometimes, from generation to generation (if present in germ cells) as the abnormal DNA is replicated time and time again. If the error blocks the expression of a gene, leads to the overexpression of a gene or leads to the production of an abnormal protein, the consequences can be severe. Such changes can even convert a normal cell to a cancerous one (Chapter 19). The results of an error in DNA can be particularly devastating when the error occurs in an essential gene present in single copy within a genome or occurs in a germ cell whose gene content is passed to the next generation. In contrast, the effects of a single abnormal RNA tend to be confined to the cell which produces it, and the effects tend to be diluted by the presence of many normal copies of the same RNA.

17.27   Most reverse transcriptases have: a) an RNA dependent DNA polymerase activity that catalyzes the production of a DNA using RNA as a template; b) an RNase activity that catalyzes the hydrolysis of the RNA in RNA:DNA hybrids; and c) a DNA dependent DNA polymerase activity which catalyzes the assembly of DNAs using DNA as a template.

17.28   RNA dependent RNA polymerases. These enzymes, which use RNAs as templates and catalyze the production of RNAs and pyrophosphate, are found in certain viruses. They use ribonucleoside triphosphates and growing RNA strands as substrates.

17.29   A reverse transcriptase can catalyze the production of one or more copies of a duplex DNA from one or more copies of an mRNA or its primary transcript. The incorporation of this duplex DNA into a genome leads to gene amplification.

17.30   For a telomerase to function, its RNA must base pair with a repeat in a single-stranded region on telomere DNA. Human telomerase RNA will only pair with human telomere repeats.

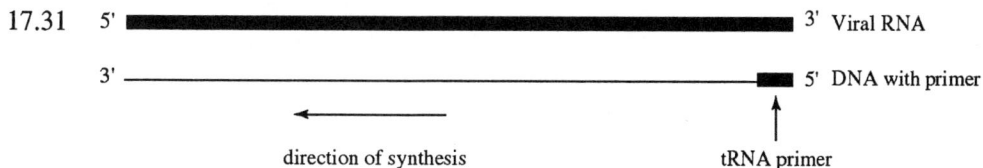

17.31

5' ████████████████████████████████████ 3' Viral RNA

3' ─────────────────────────────────── ■ 5' DNA with primer

←────────────   ↑
direction of synthesis   tRNA primer

To serve as a primer, a tRNA must be complementary to the 3' end of the viral RNA. Thus, the 3' end of the viral RNA represents a viral "gene" (although a gene is normally a segment of DNA) for the tRNA.

17.32   The error rate is greater during PCR, because PCR does not include mismatch repair systems. In addition, it is unlikely that the reaction conditions for PCR are optimal for proofreading.

17.33   No. DNA polymerases can only catalyze the 5' to 3' extension of a chain. A polymerase can not catalyze the addition of a nucleotide residue to a primer at the 5' end of the template, because this involves 3' to 5' chain extension.

17.34   Phosphate anhydride bond. Chain extension during PCR proceeds in exactly the same manner as chain extension during transcription (Figure 17.9). However, the added nucleotide residues are deoxyribonucleotide residues rather than ribonucleotide residues.

18

Protein Biosynthesis

18.1   The 16S rRNA base pairs with the Shine-Dalgarno sequence within the mRNA. This pairing helps secure the mRNA to the 30S subunit and determines which AUG is to serve as the initiation codon.

18.2   IF3: Binds to the 30S subunit and enhances the dissociation of 70S ribosomes.

IF2: A GTPase that delivers fMet-tRNA$_f^{Met}$ to the 30S:IF3 complex and monitors the formation of an fMet-tRNA$_f^{Met}$:mRNA:70S ribosome complex.

IF1: Role uncertain

18.3   The protein factors that deliver aminoacyl-tRNAs to the ribosome during the initiation and elongation processes can distinguish tRNA$_f^{Met}$ and tRNA^{Met}. IF2 is unable to bind Met-tRNA^{Met} while EF-Tu cannot bind fMet-tRNA$_f^{Met}$. The ribosome may also be involved in the division of labor between the two tRNAs, since fMet-tRNA$_f^{Met}$ binds to a codon at the P site while Met-tRNA^{Met} initially binds to a codon at the A site.

18.4   Not necessarily. The effect of compound X will depend upon how it impacts the ability of the 30S subunit to bind to 50S subunits. Compound X could, in theory, cause 30S subunits to cling more tightly to 50S subunits. In this case, the equilibrium concentration of 70S ribosomes would increase. In the case of IF3, its attachment to a 30S subunit leads to a drop in the concentration of 70S ribosomes by removing 30S subunit from the dissociation equilibrium (LeChâtelier's principle).

18.5   One. A single ribosome can be used repeatedly to retranslate the same mRNA or to translate multiple unique mRNAs. Initiation factors participate in the dissociation of ribosomes into subunits, an event that plays a key role in ribosome cycling.

18.6   H bonds, salt bridges, and hydrophobic forces are primarily responsible for maintaining the IF3:30S complex. Since complex formation is reversible and no enzyme is required, the bonds that maintain the complex are most likely noncovalent. The predominant noncovalent interactions that bind proteins (IF3 is protein) to other proteins and to RNAs (the 30S subunit is a ribonucleoprotein) are those forces listed.

18.7   The released phosphate may play a key role in maintaining an active tertiary or quaternary structure; its interactions with select amino acid side chains may be essential for native folding. Alternatively, the phosphate residue may be essential for ligand or substrate binding or for other interactions between the GTPase and its neighbors.

18.8   F2: Delivers fMet-tRNA$_f^{Met}$ to the IF3:mRNA:30S subunit complex and monitors the formation of an fMet-tRNA$_f^{Met}$:mRNA:70S ribosome complex.

EF-Tu: Delivers aminoacyl-tRNAs to the A site during polypeptide elongation and monitors codon-anticodon pairing at the A site.

EF-G: Monitors translocation and catalyzes the hydrolysis of GTP, an event that provides the energy for the conformational changes in the ribosome that drive translocation.

RF3: Stimulates the codon-specific binding of RF1 and RF2 and monitors the termination process.

A GTPase associated with SRP monitors the attachment of signal peptides and ribosomes to the ER.

18.9   Two (minimum). At least one GTP is hydrolyzed for each aminoacyl-tRNA delivered to the A site, and a second is hydrolyzed for each round of translocation. Some research suggests that the number may actually be greater than the two predicted from the information presented in this chapter. The consumed energy is used to enhance the fidelity of polypeptide assembly and to provide a thermodynamic driving force for this assembly.

18.10   N-terminal to C-terminal. Study Figures 18.6 and 18.7. The N-terminus of the growing polypeptide is transferred to each new aminoacyl-tRNA that arrives at the A site. Thus, each incoming amino acid residue becomes the C-terminus of the peptide residue attached to the tRNA at the A site.

18.11   No, at least not in theory. In reality, a messenger's life span limits the number of times it can be translated. Messenger-RNAs are degraded by nucleases that determine their half-lives (Sections 16.5 and 14.1). In some instances, the accumulation of damage may make an mRNA nontranslatable before it is degraded.

18.12   No. IF3 acts stoichiometrically, not catalytically, by removing some 30S subunits from the 70S $\rightleftarrows$ 30S + 50S equilibrium. Restated, IF3-binding temporarily destroys the equilibrium and leads to a net forward reaction, but it does not catalyze any chemical transformations.

18.13   Yes (in most cases, at least). Since the translated region is unchanged, the same polypeptide would most likely be assembled. However, there is a chance that the intron could lead to a recoding of the genetic code. The structural features within an mRNA that lead to recoding are still uncertain.

18.14   fMet-Pro-Gly-Leu-Ile-Cys

ACG

5'----AUGCCAGGGUUGAUAUGCCUUAAA---3'

18.15   Ester link. Study Figure 18.6 and the answer to Problem 18.13.

18.16   a) The codon at the A site shifts to the P site along with the peptidyl-tRNA to which it is paired, b) the empty tRNA originally at the P site is displaced from the ribosome and c) a new codon is positioned on the A site. GTP is hydrolyzed in the process.

Although not discussed in the text, there is some evidence that the empty tRNA at the P site shifts to an exit site (E site) during translocation. This empty tRNA may not exit the ribosome until an aminoacyl-tRNA pairs with the new codon positioned at the A site during translocation.

18.17   The minimum number of required tRNAs is determined by what specific leucine codon(s) appear within the fragment of mRNA being translated. If the three leucine codons are identical or if they can all be decoded through "wobble" pairing with the same Leu-tRNA^{Leu}, only 6 tRNAs are required. The same tRNA^{Leu} can be recycled and used to decode each leucine codon. Since there is a solitary codon for methionine, a single tRNA^{Met} can decode each of the two methionine codons.

18.18   Formylase (catalyzes the addition of a formyl group to Met-tRNA$_f^{Met}$), mRNA, tRNAs, aminoacyl-tRNA synthetases (catalyze the attachment of amino acids to tRNAs), ribosomes, initiation factors, elongation factors, and release factors.

18.19   In *E. coli*, amino acids are joined at a rate of up to 18 per second. Since it takes 3 nucleotide residues to encode each amino acid, the mRNA must be extended at three times this rate (roughly 54 nucleotides per second) in order for transcription to keep up with translation. Experimentally, transcription normally does proceed at close to 3 times the rate of translation.

18.20   RF1 and RF2, two codon-specific protein release factors, recognize and bind termination codons. Once bound to a termination codon, they lead to the hydrolysis of the ester bond between the completed polypeptide and the tRNA at the P site and the subsequent dissociation of the remaining translational complex which contains an mRNA, a tRNA, a ribosome and release factors.

18.21   None. The poly A tail is outside of the translated region of the messenger; it does not bind to ribosomes during the decoding process.

18.22   10-formyltetrahydrofolate plus formylase (in order to add the formyl group to Met-tRNA$_f^{Met}$), 20 protein amino acids, 20 aminoacyl-tRNA synthetases, tRNA$_f^{Met}$ plus other tRNAs capable of decoding all 61 translatable codons, mRNA (for desired polypeptide), 3 initiation factors, 3 elongation factors, 3 termination factors, 55 ribosomal proteins, 3 ribosomal rRNAs, ATP, GTP, and Mg^{2+}. Normally, purified intact ribosomes are added rather than the separate proteins and rRNAs from which they are constructed.

18.23   The mutant mRNA can be translated, since the initiation and elongation processes will proceed in a normal fashion. However, the reading of the mRNA will continue past the usual stop site and generate a polypeptide longer than the one usually assembled. Translation will stop at the next in-phase termination codon encountered (if there is one). The N-terminal portion of the longer polypeptide will be identical to the polypeptide normally synthesized. If no termination codon is encountered, no specific mechanism will exist for the release of the newly synthesized polypeptide from the tRNA to which it is attached when the translational process stops somewhere near the end of the mRNA.

18.24   Prokaryotes normally control gene expression by regulating transcription. Once a prokaryotic mRNA has been synthesized, it is normally translated immediately. The polypeptide produced is usually converted to an active form immediately (if it is initially inactive). In eukaryotes, the major control point is commonly post-transcriptional; the rate-limiting step in gene expression may occur during mRNA precursor processing, translation, or polypeptide processing.

18.25   Inactive protein precursors allow a cell to generate immediately an active polypeptide when and where it is needed.

18.26   Hydrolase. The formyl group is bound to the polypeptide through an amide bond (Figure 18.6) that is hydrolyzed during the removal of the formyl group.

18.27   a) Signal peptide: plays the central role in the attachment of a ribosome to the ER.

   b) SRP: binds a signal peptide and delivers the ribosome complex which synthesized that peptide to a ribosome binding site on the cytosolic side of the ER; it attaches to an SRP receptor on the ER during this process. SRP also monitors the attachment of the ribosome complex to the ER.

   c) Peptide translocational complex: helps transport the growing polypeptide chain through the ER membrane.

   d) GTP: powers the molecular switch (GTPase) in the SRP that monitors the attachment of the signal peptide-containing ribosome complex to the ER.

19

Genetic Diseases and Genetic Engineering

19.1   No. Based on the traditional definition, a genetic disease is one that is inherited. Unless one considers a virus infection to be an acquired genetic disease, a genetic disease cannot be acquired after the formation of a zygote (fertilized egg cell). A mutation immediately after the merger of an egg and sperm cell can lead to a disease similar to a genetic disease, but the disease is not inherited.

19.2   False (on the basis of the traditional definition of a genetic disease). A contagious disease can be passed from one adult to another and an infectious agent is involved. Defective genes are only passed from person to person at the time egg and sperm cells combine. There is an exception to this generalization if one classifies a virus infection as an acquired genetic disease. Such a classification scheme makes some genetic diseases contagious.

19.3   Yes. There is a chance that novel disease-causing mutations or combinations of mutations will arise in the future. These mutations will have to arise in egg cells, sperm cells, or their stem cell precursors in order for the mutations to become inheritable.

19.4   All of their children will have the disease if either or both partners possess two copies of the abnormal gene (assuming that a single gene is involved). However, both partners may have one normal and one abnormal gene. In this case, there is normally a 25% probability that any child will inherit two normal copies of the gene and not be a victim of the disease.

19.5   Individuals with certain genetic diseases are infertile or die prior to puberty.

19.6
$$5' \text{ CCGCGACGGCGCC } 3'$$
$$3' \text{ GGCGCTGCCGCGG } 5'$$
Normal

↓ 1st round of replication

5' CCGCGACGGCGCC 3'   +   5' CCGCGACGGCGCC 3'
3' GGCGCCGCCGCGG 5'   3' GGCGCTGCCGCGG 5'
Abnormal   Normal

↓ 2nd round of replication

5' CCGCGACGGCGCC 3'   5' CCGCGACGGCGCC 3'
3' GGCGCTGCCGCGG 5'   3' GGCGCTGCCGCGG 5'
Normal   Normal
+   +
5' CCGCGGCGGCGCC 3'   5' CCGCGACGGCGCC 3'
3' GGCGCCGCCGCGG 5'   3' GGCGCTGCCGCGG 5'
Abnormal   Normal

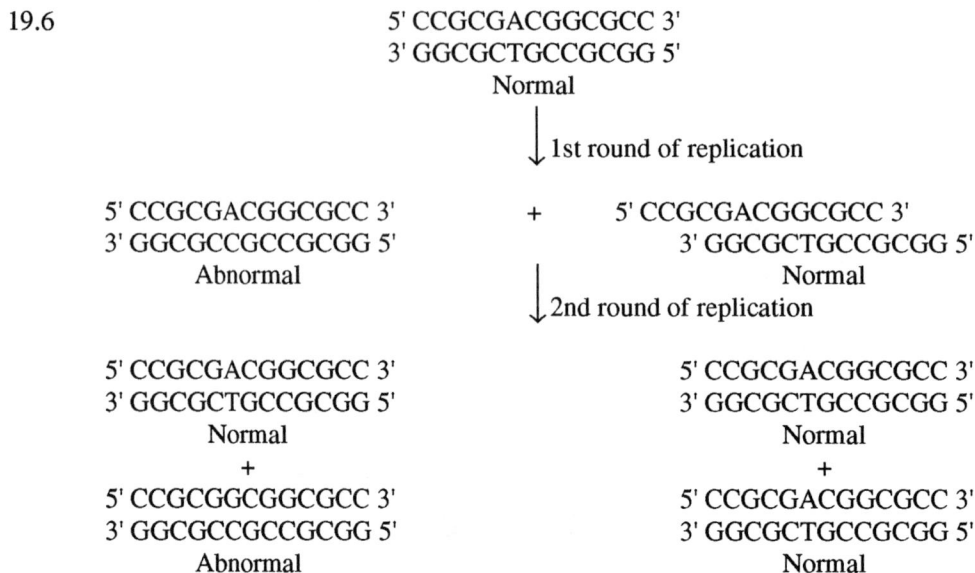

19.7   Dominantly inherited. Just one copy of the abnormal gene is all that is required to produce the abnormal enzyme responsible for the disease-linked methylation.

19.8   Hemochromatosis, if detected early, can be treated by controlling the dietary intake of iron. Hemochromatosis can also be treated with chemicals that inhibit the absorption of iron or enhance its elimination from the body.

19.9   True. The higher the mutation rate in an organism, the greater the probability that an egg or sperm cell will acquire a mutation that leads to a genetic disease. Mutation rate increases as efficiency of DNA repair decreases.

19.10   The gonads are the sites of egg and sperm cell production. Only mutations in egg or sperm cells can be passed to future generations.

19.11   Yes, if the virus contains one or more unique nucleotide sequence (which it does) or inserts its genome into a host chromosome to create distinctive RFLPs.

19.12 O_2 binding causes hemoglobin to change its conformation which, in turn, leads to the loss of a hydrophobic surface patch that exists on deoxyhemoglobin. Since this patch plays a central role in the aggregation process, aggregation does not occur in its absence.

19.13   a) No. The aspartic acid side chain is polar and charged. It is very similar to the glutamic acid side chain which normally occupies the #6 position of the β chain. In HbS, aggregation occurs because nonpolar valine side chains at the #6 position of β chains attach to nonpolar patches on neighboring molecules.

b) Aggregation could well occur with isoleucine and phenylalanine, two amino acids with nonpolar, hydrophobic side chains similar to the valine side chain. One cannot assume, however, that all nonpolar side chains will lead to the same aggregation phenomenon. The size and shape of the side chain influences its tendency to attach to specifically-shaped nonpolar patches on neighboring molecules and its interactions with neighboring side chains within the same polypeptide chain. Under certain circumstances, a nonpolar side chain may fold into the nonpolar core of a globular polypeptide to escape from surrounding water molecules. This leads to changes in the conformation of the polypeptide.

19.14   Less likely to sickle. Breathing pure O_2 decreases the concentration of deoxyhemoglobin by increasing the fraction of the oxygenated form (LeChâtelier's Principle). Only the deoxy form of a hemoglobin S molecule aggregates with neighbors.

19.15   One. One RFLP is linked to the normal gene and the second to the abnormal gene. DNA from a person with two normal genes yields two copies of the normal-linked RFLP. Both copies will appear in the same restriction band.

19.16   No. An animal that produced no β chains (β chain gene knocked out) would be unable to construct HbS or a similar molecule and would not develop symptoms of sickle cell anemia. An appropriate animal model must synthesize an abnormal β chain with characteristics similar to those of the β chain in HbS. In general, a knockout mutant can only mimic a human genetic disease if the disease is due to the lack of a gene product, an inactive gene product, or a product with greatly reduced activity.

19.17   25% probability (normally). More likely than not, both partners have sickle cell trait and carry one normal β chain gene and one abnormal β chain gene in each egg and sperm cell. These gene are randomly passed to the next generation. There is a certain very low probability that one or both parents are normal (contain no inherited abnormal β chain genes), but passed to the first child a mutated germ cell that contained an abnormal β chain gene. In this case, the second child may be at no significant risk.

19.18   Yes. PKU is a consequence of a defective gene for a single enzyme. Although one is not given enough information to determine whether or not the defect is linked to a single amino acid substitution in a polypeptide chain within the enzyme, this possibility does exist.

19.19   No. The β chain gene is virtually silent (inactive) within a fetus. A fetus primarily synthesizes HbF which contains two α chains and two γ chains, but no β chains (Section 4.9).

19.20   True. A second point mutation (change in single base pair) can restore the nucleotide sequence present in a DNA before it acquired an initial point mutation. If one mutation changes an A:T pair to a G:C pair, for example, a second mutation can change the G:C pair back to an A:T pair. However, the chance that more or less random mutations in the human genome (roughly 3×10^9 base pairs) will lead to two point mutations at exactly the same site is extremely small.

19.21   The better one's DNA repair capabilities, the lower the mutation rate and the less likely a protooncogene will be converted to an oncogene and the less likely a tumor suppressor gene will be inactivated.

19.22   False. Products of normal tumor suppressor genes help an individual avoid cancer by suppressing cell growth and division. An overactive gene product would be expected to further suppress cancer risk. However, the overactive gene product could inhibit normal and essential cell division and lead to serious health problems. There is a fine line between too little cell growth and division and too much cell growth and division.

19.23   It is likely that an oncogene would lead to abnormal embryo development and to embryo death. If true, no one will ever be born with an inherited oncogene.

19.24   Two of many possible answers follow:

5' ATCAATTGAT 3'
3' TAGTTAACTA 5'

5' CACAGCTGTG 3'
3' GTGTCGACAC 5'

19.25   "Palindrome"—a word, verse or sentence which reads the same backward and forward. Examples: madam, pop, noon, otto.

19.26   True. With an odd number of base pairs it is impossible for the 5' to 3' sequence in one strand to be the same as the 5' to 3' sequence in the complementary strand. To understand why this is the case, try to create an exception and carefully study your examples.

19.27   Plasmid Y lacks the palindrome recognized by EcoRI while Plasmid X contains at least one of these recognition and cleavage sites.

19.28   a) 14 FRAGMENTS
        Let x = Eco RI sites and o = Pst I sites

        —o   x   x   x   o   x   o   x   x   x   o   x   x—

        If one cuts a string at 13 sites, one obtains 14 pieces. Study the example given above.

        b) Digestion with Pst I alone yields 5 fragments, which, more likely than not, each contains a unique number of base pairs and appears in a separate band on a DNA fingerprint.

        c) The number of bands will be less than 5 if two or more of the restriction fragments contain the same number of base pairs.

19.29   No. If the genetic code were altered so that one or more codons specified different amino acids in bacteria and humans, the polypeptide synthesized in bacteria would contain an abnormal (noninsulin) amino acid sequence. It is highly unlikely that the abnormal protein would have any insulin-like activity.

19.30   There are an enormous number of beneficial traits that could be programmed into plants. Some examples follow:

Taste better

Grow faster

Resist physical damage (by wind, hail, harvesting machinery, etc.)

Rot more rapidly after death

Concentrate valuable minerals from the soil

Remove toxins from the soil

19.31   (See Table 19.5)

19.32   Since bacterial transcriptases are unable to recognize and bind human promoters, they cannot transcribe genes coupled to these promoters. The simplest solution? Replace the human promoter with a bacterial promoter.

19.33   The random insertion of a harmless gene can disrupt and inactivate an essential gene at the site of insertion. Designer knockout mutants are created with the selective insertion of DNA into genes (Section 19.13). Random insertions can do the same, but in a nonselective manner.

19.34   It is highly unlikely that hGH could impact growth in mice unless the mice contained growth hormone receptors able to bind this human protein. The presence of growth hormone receptors indicates that mice produce their own growth hormones. Since the human hormone binds to mouse receptors, it is mistaken (by the receptors) for the mouse hormone. Receptors are normally highly selective, so the hGH and mGH must have very similar structures.

19.35   The egg or sperm cells, because these cells carry the genetic information passed to future generations.

19.36   Three genes minimum; at least one to encode each of the three enzymes that participate in the "M" to "N" transformation. If an enzyme possesses multiple polypeptide chains (has quaternary structure), information for its production may be encoded in multiple genes. This is the case when there are two or more unique polypeptide chains within the quaternary structure.

TO THE OWNER OF THIS BOOK:

We hope that you have found *Solutions Manual for Biochemistry: A Foundation* useful. So that this book can be improved in a future edition, would you take the time to complete this sheet and return it? Thank you.

School and address: _____

Department: _____

Instructor's name: _____

1. What I like most about this book is: _____

_____

_____

2. What I like least about this book is: _____

_____

_____

3. My general reaction to this book is: _____

_____

4. The name of the course in which I used this book is: _____

_____

5. Were all of the chapters of the book assigned for you to read? _____

  If not, which ones weren't? _____

6. In the space below, or on a separate sheet of paper, please write specific suggestions for improving this book and anything else you'd care to share about your experience in using the book.

_____

_____

_____

_____

_____

Optional:

Your name: _____ Date: _____

May Brooks/Cole quote you, either in promotion for *Solutions Manual for Biochemistry: A Foundation* or in future publishing ventures?

Yes: _____ No: _____

Sincerely,

Peck Ritter

FOLD HERE

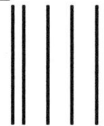

BUSINESS REPLY MAIL
FIRST CLASS   PERMIT NO. 358   PACIFIC GROVE, CA

POSTAGE WILL BE PAID BY ADDRESSEE

ATT: *Peck Ritter* _____

Brooks/Cole Publishing Company
511 Forest Lodge Road
Pacific Grove, California 93950-9968

FOLD HERE